Discoveries in the Garden

A song sparrow sings exuberantly from his perch in the lettuce patch. He shares his dominion with a scarites ground beetle, a ladybird beetle, an assassin bug that has subdued an army-worm caterpillar, and an alfalfa butterfly that rests on a nearby lettuce leaf.

Discoveries
in the Garden

James Nardi

The University of Chicago Press
Chicago and London

The University of Chicago Press, Chicago 60637
The University of Chicago Press, Ltd., London
© 2018 by James Nardi
Published 2018
Printed in the United States of America

27 26 25 24 23 22 21 20 19 18   1 2 3 4 5

ISBN-13: 978-0-226-53152-6 (cloth)
ISBN-13: 978-0-226-53166-3 (paper)
ISBN-13: 978-0-226-53183-0 (e-book)
DOI: 10.7208/chicago/9780226531830.001.0001

On the cover: Among the tendrils, pods, and flowers of pea plants, a bumble
bee searching for pollen flies over the heads of a Carolina wren and a mouse. A
ladybird beetle and its larva that are searching for aphids and thrips rest on a leaf
at lower left. A striped cucumber beetle has settled on a leaf to the right of the
ladybird beetle.

Library of Congress Cataloging-in-Publication Data

Names: Nardi, James B., 1948– author.
Title: Discoveries in the garden / James Nardi.
Description: Chicago : The University of Chicago Press, 2018. | Includes
    bibliographical references and index.
Identifiers: LCCN 2017028667 | ISBN 9780226531526 (cloth : alk. paper) |
    ISBN 9780226531663 (pbk. : alk. paper) | ISBN 9780226531830 (e-book)
Subjects: LCSH: Gardening. | Horticulture.
Classification: LCC SB450.97 .N37 2018 | DDC 635—dc23
LC record available at https://lccn.loc.gov/2017028667

♾ This paper meets the requirements of ANSI/ NISO Z39.48-1992 (Permanence of
Paper).

CONTENTS

Preface   *vii*

Introduction: Conversing with Plants   *1*

1   SEEDS   *16*

2   BUDS and STEMS, STEM CELLS and MERISTEMS: Growing Up, Down, and Out   *34*

3   The UNDERGROUND WORLD: Bulbs, Tubers, and Roots   *62*

4   The JOURNEY from FLOWER to FRUIT and SEED   *84*

5   ENERGY from the SUN and NUTRIENTS from the SOIL   *108*

6   MOVEMENTS of VINES and TENDRILS, LEAVES and FLOWERS   *138*

7   WISDOM of the WEEDS: Lessons in How Plants Face Adversity   *154*

8   PLANT COLORS   *178*

9   PLANT ODORS and OILS   *192*

10   FELLOW GARDENERS: Other Creatures Who Share Our Gardens   *210*

Epilogue   *235*

Appendix A. Important Chemicals in the Lives of Plants   *239*

Appendix B. List of Plants Mentioned in the Text   *243*

Glossary   *251*

Further Reading   *263*

Index   *267*

PREFACE

Remarkable events are everyday occurrences in a garden. Observations of these miraculous features of life arouse a desire to understand more about what we see. The excitement of discovery in biology arises not only from the sense of wonder that nature instills but also from the deeper appreciation derived from scientific experimentation. Experiments address certain mysteries about plants that observation alone cannot address and lead to even more questions about how plants do what they do. The joy of discovery comes from asking these questions and experiencing nature right at home—in backyards, in schoolyards, even indoors—and these activities offer not only the pleasures of watching vegetables grow from seeds to harvest but also the rewards of preparing and sharing the harvest in the kitchen.

Our associations with plants help maintain and restore the fragile bond between people and the natural world. Not only do plants feed our curiosity, nourish our physical bodies, and appeal to our sense of beauty, but they are also some of our best spiritual teachers. From plants we learn to have *faith* in seeds and *hope* for a good harvest. We learn *patience* as we watch plants going about their affairs at their own pace. We learn *gratitude* for the plants that provide all the essential nutrients for our health. The fruitfulness of even small gardens allows us to share an abundant harvest with others. When we grow plants, we appreciate the importance of recycling and using the natural cycles of growth and decay to keep the garden's soil, plants, and animals in harmony. We learn to appreciate contributions of these other creatures in maintaining balance and harmony in the garden. By gardening in partnership with a great diversity of creatures, we can maintain the health of the plants without resorting to the use of harmful chemicals. We feel *reverence* and awe in the company of plants, at the sight of lovely gardens, at the majesty of trees, and at the amazing abilities of plants to adjust to the world around them. As gardeners we are everyday participants in these miraculous events. Aldo Leopold reminded us, "Acts of creation are ordinarily reserved for gods and poets, but humbler folk may circumvent this restriction if they know how. To plant a pine, for example, one need be neither god nor poet; one need have only a shovel."

I am grateful to the community of family, friends, and colleagues who have helped me share these discoveries. My parents provided the opportunities and encouragement that nurtured my love of nature and gardening. Mark Bee is the talented and passionate artist of microscopy who has imaged many of the microscopic specimens. Dorothy Loudermilk and Edwin Hadley have been meticulous in their assembly and labeling of the final images. Cate Wallace of the Beckman Institute's Imaging Technology Group at the University of Illinois masterfully prepared the images of pollen that convey a hidden beauty of flowers. My friend Tony McGuigan in California has a gift for communicating his enthusiasm for gardening. His book *Habitat It and They Will Come* describes how he expresses his appreciation for fellow (but nonhuman) gardeners by creating habitats that entice them to share his garden. His ideas and suggestions have helped shape this book. From his farm in Oregon, Mark Sturges has shared his views on gardening and provided ideas for experimentation in both the garden and the kitchen. My wife, Joy, and our animal companions have shared the marvels of the garden and its earthy pleasures with me. Their eyes, noses, ears, and (in some cases) whiskers have expanded my ability to explore, discover, and appreciate. We are all grateful to gardening for being generous with its abundant gifts.

This manuscript found a welcoming home at the University of Chicago Press. As editorial director, Christie Henry offered her encouragement and support for the early, rudimentary manuscript. Miranda Martin and Christine Schwab guided the completed manuscript through its lengthy production phase. With scrupulous care and perceptive thought, Johanna Rosenbohm, in her role as copyeditor, emphasized the manuscript's better features and helped eliminate some of its worse features. To complete the publishing process, Susan Hernandez used her expertise to index the book's contents. These people made the collaborative journey from manuscript submission to book publication a joyful one.

INTRODUCTION: CONVERSING WITH PLANTS

Observe, Describe, Hypothesize

Plants are such agreeable creatures to study—we can easily raise them from seeds or from cuttings; we can place them in locations where they are easy to observe and where we can question them about what it is like to be a plant. Most of us, however, encounter plants that inhabit constantly groomed bluegrass lawns, plants that have been cut as ornamental flowers, plants that have been harvested as fruits without leaves, stems, and roots, or plants harvested as vegetables to be placed on grocery shelves. We rarely witness what goes on behind the scenes in our gardens and agricultural fields, meadows and forests. How many of us have noticed that plants are preformed inside seeds? Who among us has witnessed the journey of a colorful flower to a delicious fruit? There are plants with exploding seed capsules; plants that climb and twine, and tendrils that coil; plants with leaves and stems that move; plants that sense touch, light and dark, up and down, length of day,

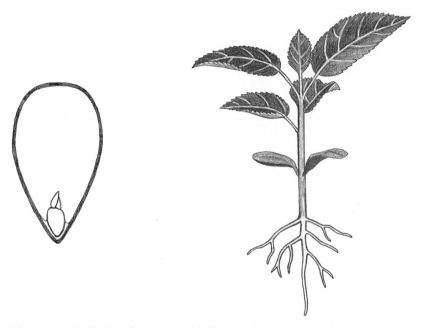

Figure I.1 An inside view of an apple seed (*left*) shows the embryonic apple tree at the tip of the seed. An apple seedling (*right*) will sprout from this seed.

and length of night. Plants may respond to the world around them in ways very different from us animals, but we soon appreciate that plants also lead lives filled with color, adventure, and amazing deeds.

From such a simple beginning as a small seed, a plant grows, extending countless roots and leaves. At a certain time, it flowers; and the flowers transform into fruits with seeds that can sprout and begin a new generation in the family of the plant. The observation that the writer Willa Cather made for trees certainly applies to all plants: "I like trees because they seem more resigned to the way they have to live than other things do." Plants accept our company, allow us to view their private lives, and adjust to the environments in which we place them. When we ask them questions about how plants do what they do and what causes them to do what they do, we wait for answers to these questions by observing how plants respond to changes in the way they have to live. By observing carefully, we can begin coming up with ideas and assumptions about how plants do what they do. When we test these assumptions by experimenting with plants, we often ask the plants questions that they probably have never been asked before.

Many secrets are still hidden among the lives of plants. For example, modern Western scientific culture has only recently learned that plants communicate with one another. Although we are just beginning to understand how plants converse with one another aboveground, much mystery enshrouds the underground conversations of plants. Without elaborate equipment but with "patience and tenacity of purpose," we can observe what has never been observed before. Our appreciation and our knowledge of how plants interact with one another and with their environments will be all the richer

Figure I.2 An apple tree has many secrets to disclose about its life aboveground and belowground.

for these discoveries. Remember the words of Henry David Thoreau as you observe and take notes: "It's not what you look at that matters, it's what you see."

When we make assumptions about phenomena that we observe in the plant world, we are posing hypotheses (*hypo* = beneath; *thesis* = rules; in other words, the rules underlying phenomena). Testing these hypotheses involves designing experiments. Experiments set out to test specific hypotheses about the lives of plants. Each hypothesis predicts an experimental result. When the prediction is tested, we see if the predicted result matches the observed result. This method of asking questions about plants, or nature in general, is known as the scientific method.

One person in particular, Francis Bacon (1561–1626), laid much of the foundation for the scientific method that we use today. While Thoreau admonished us of the importance of close observation of what we see, Bacon further admonished us to test our interpretations and assumptions of what we see. Bacon claimed that knowledge about our world is derived from our five senses, from our direct observation of nature; what we experience with our senses, however, can often be misinterpreted. We should always test our interpretations (hypotheses) and never be afraid to admit that a favorite hypothesis might be wrong. Over two hundred years later, another English scientist (Thomas Henry Huxley, 1825–1895) reminded us of this temptation to cling tenaciously to a hypothesis that simply does not match the facts that our experimental results reveal: "The great tragedy of science [is] the slaying of a beautiful hypothesis by an ugly fact."

Gardeners were some of the earliest and still are some of the most observant scientists. Many unnamed gardeners first discovered which wild plants they could cultivate, how they could enhance their harvests, and how they could promote ripening of their fruits. Over the centuries, observant gardeners have frequently challenged the generally accepted hypotheses of scientists. One such gardener was an English clergyman named Gilbert White. In *The Natural History*

Figure I.3 Honey bees are among the countless creatures whose lives intersect with the lives of plants.

of Selborne, published in 1789, White recorded what he had observed in his garden over a twenty-five-year period. While scientists and farmers of his time considered earthworms to be pests that ate seedlings and left untidy messes in the form of worm casts, Gilbert White set about testing the hypothesis that the earthworms he observed in his garden were actually benefactors of gardens. "Worms seem to be the great promoters of vegetation by boring, perforating, and loosening the soil, and rendering it pervious to rains and the fibres of plants; by drawing straws and stalks of leaves and twigs into it; and most of all, by throwing up such infinite numbers of lumps of earth called worm-casts, which, being their excrement, is a fine manure for grain and grass." Many gardeners even today are making important observations that challenge scientists to explain in greater and greater detail why and how plants do what they do.

Many of the experiments discussed in this book test the general hypothesis that gardening is simpler and more rewarding when we not only observe nature closely but also work in close partnership with nature. A popular alternative hypothesis, however, contends that successful and profitable gardening and farming require that we confront nature with synthetic pesticides, herbicides, and fertilizers. Do the facts of our discoveries in the garden help slay or support this hypothesis of conventional agriculture?

Discoveries in the Garden begins with an introduction that is followed by ten chapters, each chapter addressing a major feature of being a plant and how this feature is influenced by the plant's interactions with the world around it. Each of these ten chapters begins with an appropriate, representational illustration that highlights the particular plant feature. Each landscape scene from the garden contains plants and animals that share the garden with its vegetables and its flowers. Gardens and farms are too often portrayed as homes to homogeneous rows of plants isolated from the influence of other creatures that share their world aboveground and belowground. Instead, a garden should be thought of as a community of creatures — plant, animal, fungus, microbe — living in harmony with one another, both in spite of and because of their diverse backgrounds and activities.

Each chapter presents background information about the topic of the chapter and includes activities — observations and testing of hypotheses — that offer firsthand encounters with living plants. Each of the first nine chapters addresses how plants go about their lives, from their beginnings as seeds to their deaths as creatures that have lived full, good lives — having grown, flowered, set seeds, produced fruits, survived threats from weather and attacks from other creatures, and having accomplished all these feats in ways that we try to understand with our observations and experiments. Microscopic images of plant cells and tissues contribute an appreciation for how

Figure I.4 Other members of plant communities share the energy of sunlight captured by the plants and the nutrients they take up from the soil.

multitudes of cells arrange to produce the intricate forms of leaves, flowers, and fruits and how these arrangements of cells are responsible for the functions that plant tissues perform. Images of plant cells and tissues offer the additional benefit of helping the reader interpret how invisible changes in cells underlie the overt changes in color and form that plants can undergo. During life, these plants are nourished by nutrients recycled from plants that grew before them; after death they return these borrowed nutrients to the soil to nourish new generations of plants. These nine chapters cover the daily lives of plants in our gardens and why these ordinary lives are noteworthy and remarkable; the last chapter is devoted to our fellow animal and microbial gardeners, the creatures that share gardens with us.

Supplementing our current understanding of plant lives by suggesting observations to make and hypotheses to test offers firsthand encounters in the garden that make our interactions with members of the plant kingdom all the richer. Communicating the excitement of discovery in biology couples the sense of wonder that comes from a love of gardening and nature with the even deeper appreciation derived from scientific experimentation. Our current knowledge about plants has accumulated through such observations and testing of hypotheses.

One of the best ways to appreciate the paths we have ascended to our present level of botanical knowledge is to retrace the observations, the ideas, and the experiments—both their successes and their failures—that have guided us on our climb to each new level of knowledge. From each new height, we can gaze ahead with a clearer vision of what questions to ask and how to ask them. The paths we follow in our quest for new knowledge often twist and turn as we ascend to higher levels. As we continue to ask questions of plants, our current knowledge not only grows but also is often revised; and some information may even turn out to be entirely wrong. What we currently know about living creatures should not be expressed as rigid, immutable facts but as knowledge that is constantly advanced, refined, and often altered by new observations and experiments. The observations and hypotheses that are presented on these pages offer readers examples of the paths that have been followed by those who have experienced the excitement of discovering more about the lives of the plants who share our world. Anyone can think like a scientist, and anyone—if they observe patiently and thoughtfully—can experience the delight of new discoveries and contribute to our ever-growing body of botanical knowledge.

For those more ambitious gardeners and researchers who would like to have firsthand encounters with plants, a few easily obtained items such as seeds, pots, dishes, vials, slides, and other inexpensive supplies are needed to observe plants and to test hypotheses about their lives. Certain vegetables and fruits are readily available at grocery stores, at farmers' markets, or from local gardens. The timeless appeal of these projects lies in their simplicity and in their pointing out the overlooked but extraordinary features of "ordinary" plants.

Close observation of even the ordinary holds the promise of new discoveries to be made, as the naturalist John Burroughs proclaimed: "To find new things, take the path you took yesterday." *Discoveries in the Garden* is a science book whose projects are appropriate, engaging, and accessible for gardeners, children, and teachers, and their classes anywhere.

A Closer Look at Plants

Lives of plants are similar to our own lives in many ways, but they are very different in other ways. All creatures—small and large, animal, plant, fungus, microbe—are made up of basic units called cells. A single leaf the size of a bean or pepper leaf is made up of about fifty million such cells; a single tree the size of an apple tree contains about twenty-five trillion (25,000,000,000,000) cells. And within each of these cells lies the hereditary material known as DNA (deoxyribonucleic acid) that contains all the encoded information that a cell needs to survive and reproduce. Each cell of any creature is therefore the smallest unit with the ability to survive and reproduce on its own. Each plant cell is separated from its environment and other cells by a membrane and a cell wall. Surrounding their delicate cell membranes, plant cells have rigid cell walls that prevent them from creeping and moving about as do many animal and microbial cells. But the sturdy cell walls allow them to take up water and swell without bursting, as cells without sturdy walls would do. Plants do not have legs or wings, fins, or feet; but this movement of water into and out of plant cells in response to environmental changes nevertheless enables them to grow and move their flowers and leaves, their shoots and roots. What transpires between the initial environmental events and these final movements of plant parts in response to movements of water in and out of cells, we now know is orchestrated by simple chemical signals—the plant hormones—that pass from cell to cell.

Fitting fifty million cells, all approximately the same size, into a single leaf measuring about two inches by two inches or fifty milli-meters by fifty millimeters, means that individual cells of such a leaf are only visible under a microscope. Rather than being measured in multiple inches or millimeters, a single cell measures only a small fraction of an inch or a fraction of a millimeter (mm). Each milli-meter is divided into one thousand micrometers (μm), and cell di-mensions are usually measured in units of micrometers. The scale bars used to indicate the dimensions for most microscopic images are the width of a human hair = 100 μm = 0.1 mm = 1/250 inch. All other magnifications will be expressed as either a fraction or a mul-tiple of the width of a hair. Most plant cells—in leaves, stems, roots, or seeds—have dimensions measuring between 1/20 of a hair (5 μm) and 1/5 of a hair (20 μm).

All creatures need energy and nutrients for survival and growth. Animals eat other creatures for nutrients, energy, and survival. Rather than obtaining energy and nutrients from food that is eaten, plant cells take their nutrients directly from the soil and produce their own energy for growth by converting energy from the sun into the chemical energy of sugars. Plants obtain energy for their growth by "eating sunlight," as the botanist Tim Plowman so perceptively noted.

Within each plant cell are a number of organelles (*organ* = organ; *elle* = little). Usually the most conspicuous is the nucleus that con-tains the hereditary material of the cell. The organelles called chlo-roplasts (*chloro* = green; *plast* = form) contain countless molecules of the green pigment chlorophyll (*chloro* = green; *phyll* = leaf) in their membranes that capture the energy of sunlight. Chlorophyll passes this energy on to other molecules in the chloroplast that use the energy to produce sugars by the process known as photosynthe-sis (*photo* = light; *syn* = together; *thesis* = an arranging). Many chlo-rophyll molecules share their chloroplasts with other orange and yellow pigments, called carotenoids; like chlorophyll, these pig-ments are not soluble in water and end up in the water-repellent membranes of chloroplasts. By contrast, the water-soluble red and

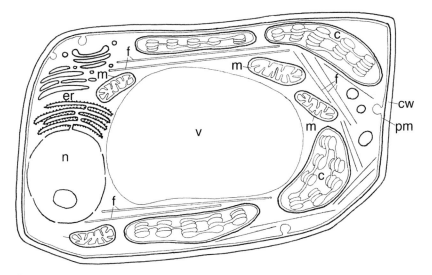

Figure I.5 This diagram of a generalized plant cell shows features shared by living cells. A plant cell, like cells of all other multicellular creatures, has a nucleus (n) associated with an endoplasmic reticulum (er), where proteins are synthesized; several mitochondria (m); filaments and tubules that form its internal cytoskeleton (f); and a plasma membrane (pm) that envelops all these organelles. A plant cell's rigid outer cell wall (cw) contributes to its immobility and sturdiness. Chloroplasts (c) that capture the energy of sunlight are found only in plant cells. The internal cytoskeleton (f) helps direct movements of organelles around the cell. Along with the cytoskeleton, the water pressure (turgor pressure) within the vacuole (v) helps maintain the form of the cell.

blue pigments of plants discussed in chapter 8—the anthocyanins and betalains—are localized to water-filled vacuoles that control the release and uptake of water from the cell. The mitochondria (*mitos* = thread; *chondrion* = granule) are organelles that use the chemical energy generated by chloroplasts to supply energy in the form of adenosine triphosphate (ATP), the universal energy currency of cells. Throughout the interior of the cell lies its skeletal framework (cytoskeleton) made up of filaments and tubules that impart a rigidity and support to the entire cell. The filaments of the cytoskeleton provide innumerable tracks along which organelles such as chloroplasts can travel within a cell.

Plants obtain essential nutrients directly from the soil and air.

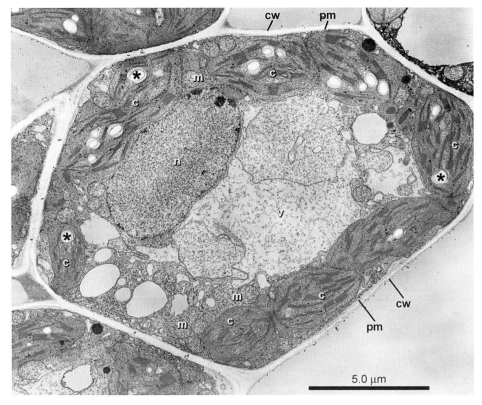

Figure I.6 In this photograph of a cell in a catnip leaf taken with an electron microscope, many of the organelles depicted in figure I.5 are labeled here with the same letters; only filaments of the cytoskeleton and the endoplasmic reticulum are not evident at this magnification. Some of the chlorophyll-containing chloroplasts are undergoing transitions to starch-containing amyloplasts (marked with *) mentioned in chapters 1 and 3.

The supply of nutrients in the soil is constantly renewed by the action of creatures belowground known as decomposers that recycle nutrients from the remains of everyone and everything that once used these same nutrients while they were alive. These decomposers assure that the growth of living creatures is balanced by the recycling of remains of once-living creatures. In the garden and wherever nature is left to its own devices, death is but the prelude to rebirth.

In addition to producing essential compounds such as hormones, amino acids, nucleic acids, and sugars that are required for their development and reproduction, plants produce thousands of other chemical compounds that are not essential for their survival but certainly influence their interactions with their environment and with other plants. These are referred to as secondary metabolites. Some of these are health-promoting compounds, such as antioxidants found in fruits and vegetables. Others are responsible for the myriad pleasant, enticing colors, odors and flavors of flowers, fruits, and herbs. The sweet scents of flowers, the fragrances of mints, and the appeal of coffee and chocolate can be attributed to a variety of secondary metabolites. Many have proved to be important in medicine, and others are being investigated for their antimicrobial and anticancer properties. Many function as repellents or toxins for insects and other plant-feeding animals or as inhibitors of the germination and growth of competing plants. Some plant compounds are so versatile that they can serve in multiple roles, depending on whom or what they encounter.

These special features of plants certainly have a great influence on how plants respond to their environment and how plants carry out their everyday tasks of growing—obtaining nutrients from the soil, gathering energy from the sun, producing substances that only plants can produce, and finally, aging like the rest of us.

Using plants as subjects of observation and experimentation makes the excitement of scientific discovery accessible to a wide audience. In addition to engaging and inspiring people to understand why plants do what they do, many images in *Discoveries in the Garden* present the rich biodiversity and aesthetics of the natural world, from macroscopic to microscopic. As a quote from the philosopher Simone Weil reminds us, "The true definition of science is this: the study of the beauty of the world." Using the microscope to expand our understanding of how and why plants do what they do reveals an inner beauty of plants that enhances our appreciation of their outer beauty.

Appendix A at the end of the book presents the chemical structures of representative plant hormones, plant pigments, and plant secondary metabolites that shape the lives of all plants. Additionally, many vegetables, trees, flowers and weeds are discussed on these pages. Thus appendix B is devoted not only to providing both the common names and scientific names of these plants but also to indicating the family association of each plant (and listing other examples of plants within families that are named in the text). In this appendix, plants are arranged alphabetically by scientific family name; within each family, plants are alphabetically listed by their common names. The genus and species of each plant follow its common name.

1

SEEDS

I have great faith in a seed. Convince me that you have a seed there, and I am prepared to expect wonders.

HENRY DAVID THOREAU

A seed is a marvel of creation; from so simple a beginning and with only energy from the sun and nutrients from air and soil, each seed transforms into a complete plant with leaves, stems, roots, flowers, fruits, and seeds of its own. As you can see if you open a seed such as a bean seed, a miniature version—the embryo—of the future plant is preformed in this small parcel of just a few cells. These cells of the seed are destined to form all future plant parts; through their many divisions during the plant's life, the basic form of each plant contained in the seed grows and

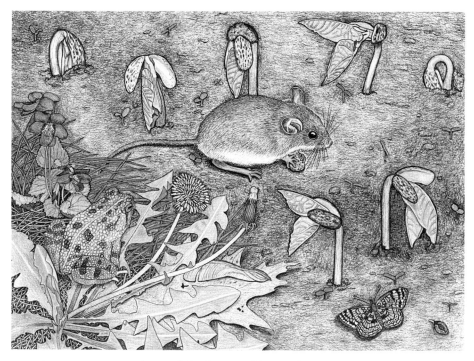

Figure 1.1 A mouse samples a cotyledon from one of the many bean seeds that are germinating in this newly planted garden. A toad watches from the patch of dandelions, violets, and blue-grass. A pearl crescent butterfly spreads its wings in the warm May sunshine.

matures. What is also remarkable about the transformation of a seed to a plant is that this basic form of the plant is retained throughout all the growth that takes place. Growth is never haphazard and disorganized but orchestrated by groups of cells at root tips and stem tips, where growth and cell division are concentrated. As the plant grows, these cells will continue to divide to produce more cells like themselves but also cells that become specialized leaf cells, root cells, flower cells. The dividing cells remain in specific places throughout the plant called buds and meristems. Each seed carries these untold promises and possibilities. A Welsh proverb well expresses what wonders can unfold when a seed sets forth on its journey of germination and growth: "A seed hidden in the heart of an apple is an orchard invisible."

The Future Plant Is Preformed in Its Seed

OBSERVE: The most conspicuous features of a germinating bean seed or sunflower seed are the seed leaves, or cotyledons (*cotyle* = cup-shaped); these provide the first nutrients for the baby plant or embryo and are the first part of the future plant to disappear as the nutrients they contain are transferred to the growing seedling. Each germinating seedling sends one growing tip into the soil below the cotyledons—the future root or hypocotyl (*hypo* = below; *cotyle* = cotyledon)—and the other growing tip into the sky above the cotyledons, the future stem, or epicotyl (*epi* = above; *cotyle* = cotyledon). Split almost any seed that you have soaked in water for several hours along its plane of least resistance, and you will reveal the future plant or embryo hidden within (fig. 1.2).

Seeds of flowering plants begin their sheltered lives within flowers that then transform into fruits as the seeds mature. All flowering plants, or angiosperms (*angio* = enclosed; *sperm* = seed) form fruits with enclosed seeds that have either one cotyledon or two cotyledons—no more, no less. The cotyledons of these seeds are an important feature that distinguishes the two major lineages of the flowering plants. Of the 235,000 species of angiosperms, 65,000 species have seeds with one cotyledon; these plants include corn, wheat, oats and all grasses, asparagus, onions, lilies, irises, palms, and orchids. These are referred to as monocots (*mono* = one; *cot* = abbreviation for cotyledon). Melons, beans, tomatoes, cabbage, and carrots are among the remaining 170,000 species whose seeds have two cotyledons; these are the dicots (*di* = two; *cot* = abbreviation for cotyledon).

Evergreen conifers—pines, spruces, firs—also have seeds with epicotyls, hypocotyls, and cotyledons that are surrounded by nutritive tissue. Their seeds, however, do not lie within protective fruits,

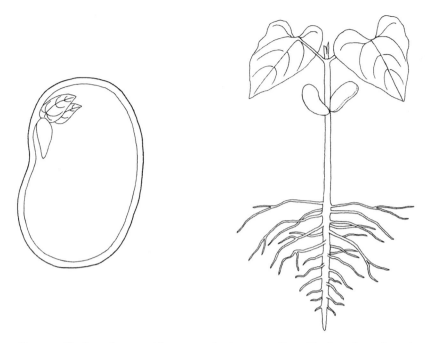

Figure 1.2 The future leaves and future root of a plant are preformed in the embryo of a seed.

but instead lie exposed on the surfaces of their cone scales. For this reason they are referred to as gymnosperms (*gymnos* = naked; *sperm* = seed). All living gymnosperms—conifers and their other relatives with naked seeds—make up only 720 species worldwide, representing a mere 0.3 percent of all plants with seeds. By producing both seeds and spores, angiosperms and gymnosperms are distinguished from other green plants such as mosses, ferns, and algae that produce only spores and have never acquired the ability to form seeds. (Additional information on seeds and spores is provided in the section of chapter 4 titled "The Difference between Seeds and Spores.")

Like the seeds of many flowering plants, the seeds of many conifers—pine nuts—are relished for their flavor and nutrition. An inside view of a pine nut reveals its resemblance to flavorful seeds of

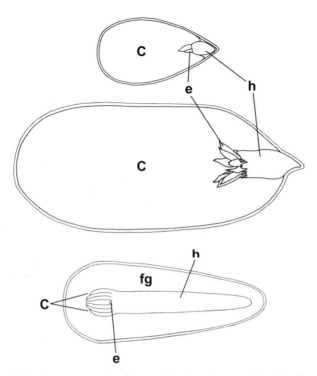

Figure 1.3 Within the mature seeds of an apple (*top*) and a peanut (*center*), their nutritive endosperms have all been transferred to the two cotyledons (c) that will nourish the early growth of the embryos that they embrace. The epicotyls (e) and hypocotyls (h) of each embryo are marked. Within the pine seed (= pine nut, *bottom*), the embryo and its cotyledons are surrounded by nutritive tissue whose origin differs from the origin of the nutritive endosperm of angiosperms. Endosperm arises from the fusion of a sperm cell with the two nuclei of a female cell, whereas the nutritive tissue of the pine seed is derived from the many divisions of a single female cell (fg = female gametophyte), without any contribution from pollen sperm. See chapter 4 for more details.

angiosperms like the peanut (fig. 1.3). The pine embryo and its cotyledons, however, are embedded in maternal nutritive tissue found only in gymnosperms.

Peanuts are relatives of beans; and whenever you munch on peanuts, you are munching mostly on cotyledons of peanut embryos. Within each peanut, the future peanut plant is nestled between the two large cotyledons at one end of the seed. Look closely and you can make out the peanut embryo's future leaves pointing inward and its future root pointing outward.

In the teardrop-shaped apple seed, the future root lies at the tapered end of the seed and points outward. On top of this future root lies a tiny group of cells destined to become the first leaves and the trunk of the apple tree. The cotyledons occupy the broad end of the apple seed and meet at the point where the future root joins the future shoot.

As they first form inside fruits, all angiosperm seeds begin life endowed with a store of nutrients in the form of a tissue called endosperm (*endo* = within; *sperm* = seed). Endosperm tissue arises at the time of pollination, at the same time that the angiosperm embryo is conceived. Unlike the maternal nutritive tissue that surrounds the pine embryo, the endosperm is both maternal and paternal in origin. Chapter 4 describes the details of how the pollination of a flower with a single pollen grain can produce not only an embryo but also its accompanying endosperm. Embryonic plants found in seeds (such as those of apples, beans, sunflowers, and peanuts) with conspicuous cotyledons transfer most of the stored nutrients from their endosperms to their cotyledons long before they germinate.

The angiosperm seeds we have examined so far have done just that; by the time they germinate, their endosperms have completely disappeared. Many seeds, however—such as tomatoes, corn, peppers, rye, and wheat—do not have such conspicuous cotyledons. These seeds have retained their original endosperms as a reservoir of nutrients that they can tap for nourishment as they germinate. Views of the interiors of a rye seed, a pepper seed, and a corn seed show these differences in endosperms and cotyledons, plus the first days of germination for different seeds (fig. 1.4).

HYPOTHESIZE: What happens if you gently remove the two cotyledons from a newly germinated bean seed? Compare the growth of

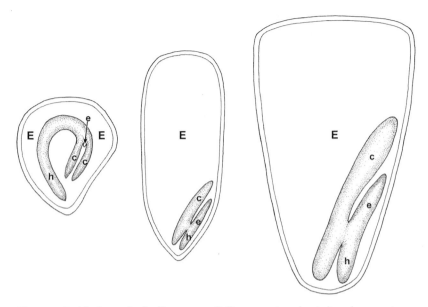

Figure 1.4 Inside the seeds of a dicot pepper (*left*), monocot rye (*center*), and monocot corn (*right*), the nutritive endosperms (E) surround the embryos, or future plants, each having one or two cotyledons (c); a future root, or hypocotyl (h); and a future stem, or epicotyl (e).

a bean seedling having both of its cotyledons with a bean seedling that has lost both or one of its cotyledons. Can whatever nutrients that are derived from cotyledons ever be replaced by nutrients from other sources? Can beans that have lost one or both cotyledons ever grow as large as beans with both cotyledons? When can cotyledons be removed from bean seedlings without having any influence on the development of the future bean plant?

How does a newly sprouted sunflower respond after five or six hours if one or both of its cotyledons are gently wrapped with foil to block light from reaching the cells of the cotyledons? What happens if the epicotyls of these newly sprouted seedlings are removed with fine scissors and only their cotyledons remain? The observed results should be less puzzling after reading about the experiments of Charles and Francis Darwin in chapter 2.

Knowing Which Way to Sprout

OBSERVE: Knowing the difference between up and down is important to all plants. In each seed we saw that the future leaves point in one direction and the future root points in the opposite direction. The future roots and future leaves have definite orientations within seeds. When seeds are carefully planted or even scattered in the garden, however, they can be placed in any position and any orientation and still manage to send their roots down and their leaves up. Every seed has a distinct sense of direction. A kernel of corn is one of the best seeds in which to demonstrate this ability of plants to distinguish up from down; the first root of corn emerges from the pointed end of the seed and the first leaf emerges from the flat end. The large seeds of corn are easy to glue on dry filter paper in different orientations (fig. 1.5). As you are sitting at your desk or table, glue some pointing toward you, some pointing to your left, some pointing to your right, and some pointing away from you. Once the glue has thoroughly dried, moisten the filter paper and cover the dish to hold the moisture for a few days. Also tape the lid to the 100-millimeter petri dish so it remains in place when you now move the dish from its horizontal position to a vertical position that germinating seeds would naturally occupy in the soil. Do the first roots from the four corn seeds all grow in the same direction?

HYPOTHESIZE: Let the first roots grow about one inch. What do you think will happen to these growing roots of corn if you rotate the petri dish 180 degrees? What happens if you place the dish in a horizontal position? In this position the roots are prevented from growing downward; they can only grow up or grow to the side.

Making sure the filter paper is moist, flip the horizontal dish upside down. The filter paper and seeds should remain on the original bottom of the petri dish; the roots and shoots of the corn seedlings

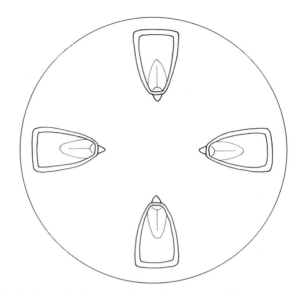

Figure 1.5 Seeds of corn glued to the surface of filter paper will begin sprouting after being moistened with water and placed in a petri dish. Mount the dish vertically and wait a couple of days for the first roots, or hypocotyls, to emerge from the tapered ends of the seeds.

now must adjust to being upside down, being able to grow downward and sideways but not upward.

How are roots able to sense the difference between up and down? Just as our own sense of up and down is conveyed by the movement of tiny granules that rub against tiny cell structures within a chamber of our inner ear, the cells of a plant's growing root tip contain small, round, dense granules called statoliths (*stato* = resting; *lith* = stone) that shift about in the cell as the cell changes position (fig. 1.6). These statoliths turn out to be dense starch granules that form in special chlorophyll-free chloroplasts called amyloplasts (*amylo* = starch; *plast* = form). Figure I.6 in the introduction represents a good example of this transformation from chloroplast to amyloplast. When the statoliths settle to the bottom of the cells, the root grows downward and growth occurs uniformly around the root tip. But if the root is placed on its side, the statoliths settle on that side of the cell. Now

root cap

statoliths

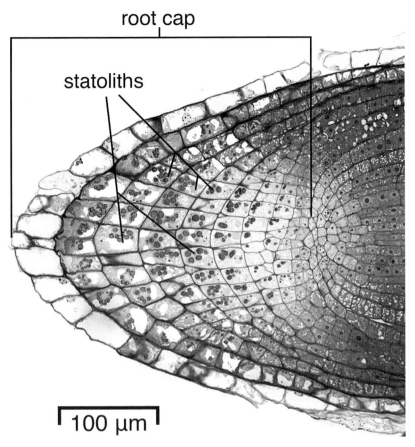

100 µm

Figure 1.6 At the very tip of the growing root, the gravity-sensing statoliths lie within those cells that make up the root cap (bracket) of this radish root. As the orientation of the root changes, the statoliths shift their positions within the cells of the root cap.

more growth occurs on the opposite side of the root, and the root starts growing down. What happens when the root tip is placed up-side down and the statoliths settle to the tops of the cells? Somehow this shift in the statoliths prompts the root tip to do an about-face and reverse the direction of its growth completely. We now under-stand that the repositioning of statoliths is translated by cells of root tips into directional root growth. The details of the events that transpire between the initial movement of statoliths and the final movement of root tips, however, are still being filled in by carefully planned hypotheses and experiments, many of which are discussed on future pages.

HYPOTHESIZE: If the movement of statoliths within the cells of the root cap conveys information to the root about its orientation in the soil, then one would hypothesize that removal or damaging of the root cap cells would alter the normal response of a root to gravity.

Repeat the previous experiment with corn seeds that have been attached to filter paper in four different orientations. After the first roots have sprouted from the corn seeds, carefully remove their root caps with fine tweezers or a fine pin. Observe the subsequent growth of these roots. How does absence of a root cap and its statoliths influence the direction in which a root grows?

Knowing When to Sprout

A seed remains at rest until it receives the signals from its environment to go into action. These signals tell the seed that conditions are right to begin growing. There is no point in sprouting if the air and soil are too cold or too dry or if you are buried too deep in the dark soil to reach the far-off light.

OBSERVE: To watch the chain of events as a seed germinates and a new plant is born, place radish seeds on the moist surface of damp filter paper placed in the bottom of a covered petri dish. Which part of the future plant emerges first from the seed? What is the fuzzy material that surrounds this first sprout? A close-up view reveals that the fuzz is not mold but represents thousands of hairs that project from the cells on the surface layer of the sprout (fig. 1.7). As this first sprout from the radish seed extends through the soil, its thousands of hairs also extend in all directions throughout the soil, into tiny pores and spaces in search of the water and nutrients essential

Figure 1.7 The countless root hairs of a radish seedling extend into the countless tiny pores in the soil that surround the germinating seed.

for its growth. They manage to reach out into the soil without getting hopelessly intertwined and tangled with one another—a feat as remarkable as the ability of the thousands of fish in a school to swim and the hundreds of birds in a flock to fly in their orderly arrays.

Seeds that have been buried deep underground can remain dormant but healthy for many years. Seeds of a wildflower from Siberia currently hold the remarkable record of existing 32,000 years in a dormant but healthy state. Russian scientists reported in 2012 that they had discovered seeds of a flower in the pink family (*Silene*) that had been stashed away in a ground squirrel's burrow 32,000 years earlier—a time when woolly mammoths shared the landscape with these flowers. Soon after its construction, the burrow had been buried under 125 feet of sediment, and the seeds remained frozen for all those years. Buried seeds face two dilemmas: lack of light energy and lack of oxygen; and they germinate only when they are brought to the surface of the soil.

HYPOTHESIZE: Do different seeds have different environmental requirements for germination? Can all—or only some—seeds germinate in the dark? Seeds that have been placed in dishes with moist filter paper do not encounter a shortage of oxygen; but these seeds may require not only moisture and oxygen but also light before they send forth their first shoot. To test the responses of seeds to the absence and presence of light, place tobacco seeds in three 100-millimeter petri dishes with moist filter paper. After adding a dusting of tiny tobacco seeds to each dish, place one dish in the light and a second dish in complete darkness for five days. Place the third dish in complete darkness for two days; expose it to light for one day; and then return it to the dark for two more days. Follow the same procedure with radish seeds. For some seeds, the presence of moisture and oxygen alone may be sufficient to trigger their germination. Can you think why different seeds may differ in their germination requirements?

Can seeds germinate at any time of year? What about seeds of desert plants that face many months of drought? Many seeds that mature at the end of summer, such as those of apples and oaks and prairie flowers, have to experience the cold days of winter before they will germinate. Rather than germinating at the beginning of winter and soon exposing themselves to freezing temperatures, seeds remain dormant and will not germinate until after they have been exposed to several weeks of temperatures below 40 degrees Fahrenheit. Seeds can experience these temperatures either during a natural winter outdoors or during an artificial winter in a refrigerator. After the passage of winter or a minimum number of days below 40 degrees Fahrenheit (5 degrees Celsius), germination of a seed can proceed without fear of its new sprouts being frost bitten.

While germination for some seeds requires exposure to a critical cold period, certain other seeds must await exposure to fire, to smoke, or simply to rain before they begin sprouting. Seeds from many tropical plants germinate only within a relatively high temperature range of 95 to 104 degrees Fahrenheit (35 to 40 degrees Celsius). Thick, hard, and impermeable seed coats are often cracked by brief exposure to fire, increasing exposure of the seed's inner embryo to moisture and air. While not necessarily essential for germination, smoke exposure of seeds from some prairie plants increases their germination by 30 to 40 percent, as researchers at Chicago Botanic Garden have demonstrated. Seeds use cues for their germination that are molded by the moisture, light, and temperature conditions that seeds must face in their particular habitats.

A factor that suppresses growth is found in many dormant seeds; this factor then disappears as seeds initiate their growth. This same factor is found in dormant, resting buds. Only with the advent of spring do levels of growth-promoting factors rise as the levels of the growth-inhibiting factor drop. We shall discover that a balance between growth-promoting factors and growth-inhibiting factors control seed germination as well as so many other events in a plant's life. In the case of seeds, one factor inhibits their germination and maintains their dormant condition until lengthening days and warmer temperatures prompt the levels of growth-promoting factors to increase.

The observation that a cold treatment was essential for the germination of certain seeds was the first step in developing the hypothesis that some chemical factor is responsible for inhibiting the germination of these seeds. A simple organic substance named abscisic acid was purified from dormant seeds. The level of growth-inhibiting abscisic acid drops as seeds begin to germinate; and the levels of growth-enhancing factor(s) begin to rise. Additional experiments have tested the hypothesis that abscisic acid inhibits not only seed germination but also bud growth; the observed results with many different seeds and many different plants match results that one would expect if this hypothesis is correct. The name abscisic acid was chosen at a time when this substance was also hypothesized to trigger the detachment and fall of aging leaves from plants (*absciss* = cut off). Closer inspection of the abscission process, however, revealed that abscisic acid does not stimulate the fall of older, mature leaves. Instead, abscisic acid inhibits the growth of the adjacent bud at the base of the old leaf. By acting in this way, abscisic acid prevents future leaves within the bud from appearing until warmer days of spring arrive.

The inhibitory influence of abscisic acid, however, is counterbalanced by the action of at least one growth-stimulating factor. A growth factor in the seed converts starch (the stored, dormant form of energy in the seed) to sugars (the active, usable form of energy for seed germination). This growth factor is a substance known as gibberellic acid, active not only in the germination of seeds but also in other events in the lives of plants (fig. 1.8). The influence of this growth factor, in fact, was first observed on rice seedlings that were growing exceptionally tall and spindly after being infected with a fungus known as *Gibberella fujikuroi*. This fungus was producing something that stimulated the cells of the rice plants to stretch and grow; that something turned out to be a simple compound that was named after the fungus from which it was first isolated. Ever since its first discovery in fungi, gibberellic acid has been seen as instrumental in the growth of plants, from their germination as seeds in the spring to their aging in the autumn. Hormone (*hormon* = arouse) is the name given to these chemical agents such as abscisic acid and gibberellic acid that "arouse" particular actions or inactions in a plant's life.

Although one of these hormones may be mainly responsible for orchestrating an important event in the life of a plant, all these hormones exist together in a plant and none of them act entirely alone. Plant hormones interact with one another—at different concentrations, at different locations, at different times—in ways we are still discovering. As other important events in the lives of plants are

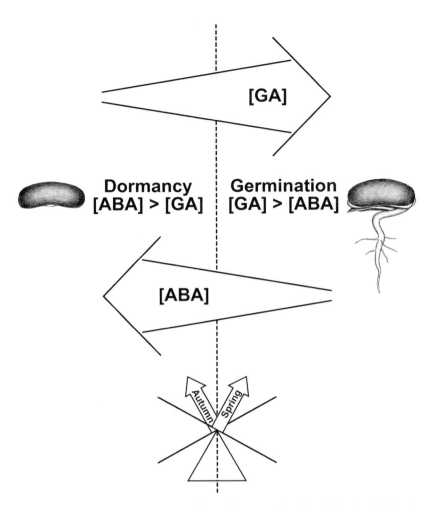

Figure 1.8 The relative concentrations of the two hormones abscisic acid (ABA) and gibberellic acid (GA) rise and fall with the seasons of the year, orchestrating important events in the life of a plant such as dormancy, seed germination, and the opening of buds. The balance of hormone levels determines the fate of buds and seeds.

discussed in the following pages, some of these same familiar plant hormones, along with several new ones, will make their appearances.

OBSERVE: Even among seeds (acorns) of two such closely related plant species as red oak and white oak, two very different strategies

Figure 1.9 The acorns of red oaks (*left*) and those of white oaks (*right*) have not only different germination requirements but also different sprouting strategies.

exist for facing the freezing days of winter. In the autumn, gather acorns of red and white oaks and lay them on the surface of moist soil in a large pot; then observe how these two different but closely related seeds anticipate the arrival of winter.

The two major groups of oaks—the red oak group and the white oak group—are easily distinguished by the distinctive shapes of their leaves and their acorns. Members of the red oak group have leaf lobes that end in tapered points; members of the white oak group have leaf lobes that end with rounded points (fig. 1.9). They also differ not only in their germination requirements but also in the time required for the maturation of their acorns—two years for red oaks but only one year for white oaks.

2

BUDS AND STEMS, STEM CELLS AND MERISTEMS: GROWING UP, DOWN, AND OUT

Each bud of a plant stem contains one or more special cells called stem cells, each of which can give rise to an entire flower, a leaf, a root, or even an entire new plant. A *stem* cell has some of the same attributes as a plant *stem*. The first, however, is a single cell, while the second is a tissue made up of many cells; they both represent fundamental, unspecialized structures from which all other plant structures such as flowers, leaves, and roots can arise. However, while each bud contains thousands of specialized cells but only one or a few stem cells, each stem of a plant can have many such buds and just as many or more stem cells. Stem cells are found in all creatures wherever unspecialized cells become specialized to perform particular tasks.

Stem cells are concentrated in parts of each plant called

Figure 2.1 Sunflowers attract not only mice and goldfinches but also pollinators such as honey bees (*upper left*), male horse flies (*center left*), and the large blue-black sphex wasps. When not pollinating flowers, the female sphex wasp searches for katydids (*lower left*), using them to provision the underground chambers she constructs for her eggs and larvae.

meristematic regions. Those regions that are involved in growing up or down form meristematic buds of shoots and roots (figs. 2.2, 2.3); those regions involved in growing out to expand the diameter of a shoot or root form meristematic rings beneath the surface and around the circumference of each shoot and root (figs. 2.4, 2.5). The Greek word *meristos* means "divisible" and refers to a feature of all meri*stem*atic regions; they represent regions that are home to stem cells where cells continually divide. Stem cells not only have the special ability to give rise to an entire line of new, specialized cells but they also divide to produce more unspecialized stem cells like themselves.

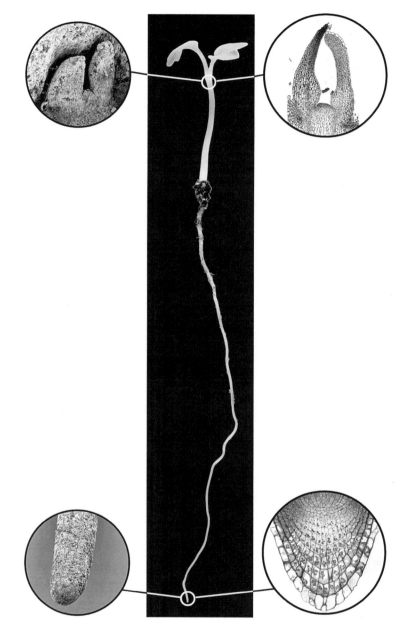

Figure 2.2 Here are close-up views of the two meristematic buds of a radish seedling—its root tip and its apical bud. Scanning electron microscope images on the left side show magnified surface views of the apical meristem and the root tip meristem. Sections through these same regions on the right side show the interior arrangement of cells within these two meristems.

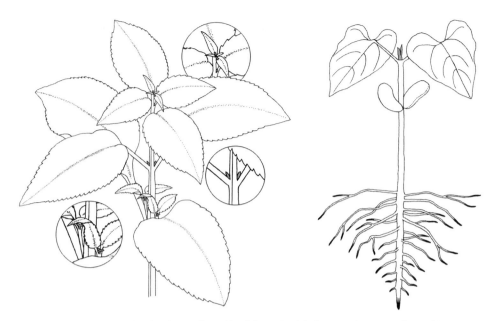

Figure 2.3 Stem cells of *Coleus* (*left*) and bush bean (*right*) shoots and roots are located in meristematic buds aboveground and meristematic root tips belowground. These meristems have been shaded and amplified.

Stems and roots of garden plants increase not only in length but also in diameter; they grow not only up and down but also out. To add cells to its diameter, every plant has a cambium—a thin ring of meristem around its trunk and beneath its surface that extends all the way from root tips belowground to buds aboveground. The Latin name *cambium* (plural = *cambia*) means "a change" and refers to the remarkable change that occurs in dividing, unspecialized cells of the meristematic cambium as they transform into specialized cells. The fate of these cells is determined by whether they lie within the ring or outside the ring of dividing cambial cells. The stem cells of the cambium divide to produce not only more unspecialized stem cells like themselves but also specialized cells toward the outer surface of the root and another class of specialized cells toward the center of the root (fig. 2.4). Stem cells within the cambial ring divide

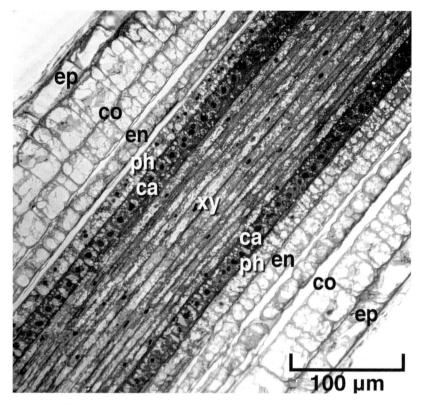

Figure 2.4 This longitudinal section of a radish seedling root spans its entire width, from outermost epidermis (ep) to central innermost xylem (xy). The dark ring of meristematic cambium (ca) separates the phloem cells (ph) outside the ring from the xylem cells (xy) inside the ring. Cells of the root cortex (co) and the endodermis (en) separate phloem cells (ph) from epidermal cells (ep).

toward the center of the plant (toward the wood) and become xylem cells (*xylo* = wood), which conduct water and minerals from below. Stem cells on the outer surface of the cambial ring divide toward the periphery of the trunk (toward the bark) and become phloem cells (*phloem* = bark), which conduct sugars formed in leaves above (fig. 2.4). The plant continually expands outward as stem cells of the cambium generate the xylem and phloem conducting cells that form its vascular (*vascu* = duct) transport system.

OBSERVE: A good way to see how plants grow out as well as up and down is to locate the very thin ring of stem cells that make up the cambium in a cross-section of a root. Place the tip of a small carrot root in a tube filled with green or blue food coloring. The cells that are specialized for conducting water and minerals from the soil will do the job of moving the blue dye from the root's tapered tip to the wide end of the root where the leaves once were. After two or three hours, wash the dye from the outside of the root tip and take a sharp knife or single-edge razor blade to cut across the root to see how far the dye has been transported along the root. The dye labels the specialized xylem cells that carry water and minerals toward the leaves, flowers, and fruits of a plant. The stem cells of the cambium form a thin ring of cells encircling the central disk of cells that contain the dye (fig. 2.5). Encircling this ring of cambium cells is a ring of phloem cells—cells specialized for transporting sugars in the opposite direction: from leaves, flowers, and fruits to the roots.

The ring of cambium stem cells produces new cells not only to expand the girth of a plant but also to replace its damaged or missing cells. The stem cells of the cambium can remarkably heal together two wounded stems and enable gardeners to graft two stems of the same plant species or even different but related species of plants (fig. 2.6). By grafting a stem of a red tomato to a yellow tomato, a mosaic tomato plant arises that produces both red tomatoes and yellow tomatoes. Since tomatoes are members of the same plant family as potatoes, grafting a tomato stem to a potato stem results in a plant that is part tomato and part potato—a plant that can bear red tomatoes aboveground and russet potatoes belowground. The secret of the graft's success lies in the ability of the stem cells from the two plants' cambia to work together to replace and repair the cells that were removed during the grafting. The two stems are held together with tape during the healing process. After a couple of weeks, the meristematic cells of the tomato cambium and the potato cambium have not only multiplied to replace missing and dead cells but have also thoroughly healed and joined the wounded surfaces of the two stems.

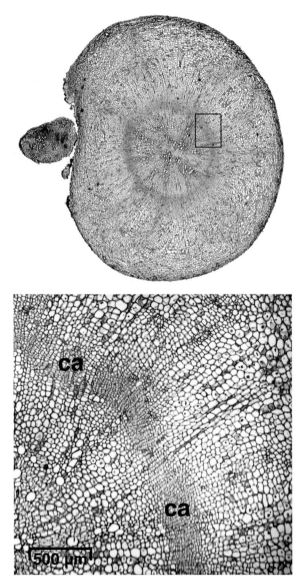

Figure 2.5 *Top:* The thin ring of cambium stem cells in a transverse section of a carrot root lies between the xylem cells in the center of the root and the phloem cells of the outer layer of the root. The rectangle spans the cambium ring with xylem cells occupying the lower left of the rectangle and phloem cells occupying the upper right. A section of a lateral root projects from the left side of the main root. *Bottom:* A close-up of the ring of cambium meristem (ca) is shown in this transverse section of this same carrot root. This region is marked by the rectangle in the figure above.

Figure 2.6 A tomato plant (*left*) grows next to a potato plant. Both are members of the same family, the Solanaceae, or nightshade family.

Place two six- to eight-inch tall plants close together in the same large pot. Gently tie the stems together about halfway between the soil and the tops of the plants with string or a twist tie (fig. 2.7).

With a single-edge razor blade, remove a slice of stem from each plant just above the place where the two stems are tied. Each slice should be about an inch long but should penetrate only as deep as a third of each stem's diameter—in other words, deep enough to reach the cambium of each stem.

The two cut surfaces should fit together snugly and should now be wrapped with tape to ensure that the plants heal and remain healthy. If the plants look healthy and possibly even show new growth after

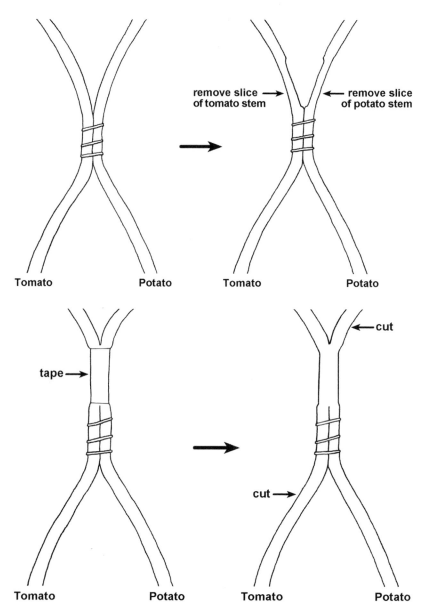

Figure 2.7 Top: The first steps in the grafting of a potato stem to a tomato stem. *Bottom:* About two weeks after joining the cut surfaces of a potato stem and tomato stem, the meristematic cells of their cambia have divided to heal the wound.

ten to fourteen days, use the single-edge razor blade to cut off the top of the potato plant an inch above the graft and cut off the lower portion of the tomato plant about an inch below the graft. After two or three more days, remove the tie and grafting tape. The mosaic tomato-potato plant will set tomatoes in the summer, and its potato harvest will be ready after the tomato stems and leaves wilt at the end of the summer.

HYPOTHESIZE: Graft a tomato plant that produces large red (LR) tomatoes and a plant that produces red cherry (RC) tomatoes. Use plants that are both six to eight inches tall and join them in the same way that you grafted the red tomato plant to the potato plant. Compare the number and size of fruit on the two different mosaic tomatoes (one having LR roots and RC shoots; the other having RC roots and LR shoots) with the number and size of the tomatoes on the ungrafted LR and RC tomato plants. Do certain combinations of roots and shoots result in higher-than-average or lower-than-average fruit harvests?

The Topmost (Apical) Bud and Apical Dominance

Buds are homes for stem cells that are constantly adding new cells and new growth to the plant. These meristematic regions communicate with one another so that growth of a stem is by no means haphazard and is orchestrated predominantly from its topmost bud. In fact, the different meristems of the plant have a well-established hierarchy, or ranking order, that the topmost bud maintains by communicating its dominance to buds lower on the stem.

Communication among cells in our own bodies can occur in three different ways: (1) Cells can move from place to place just as our blood cells constantly move throughout our bodies. (2) Cells can grow and extend in particular directions. A single nerve cell in our leg can stretch as far as from our back to a toe. (3) Communication also occurs when chemical messengers are sent forth that travel to distant locations, passing from cell to cell. These chemicals are produced at one location and then distributed to other locations.

Cells of plants with their rigid cell walls neither move nor stretch significant distances from one point to another along the stem; but they can send chemical messages, within the same plant or from one plant to a nearby plant. Chemicals that act as these messengers are known as hormones. Hormones such as gibberellic acid and abscisic acid mentioned earlier in chapter 1 can bring about changes in a plant by stimulating some actions and inhibiting others.

Plant-growing tips such as buds serve as homes to stem cells, and these buds are also homes to a hormone that orchestrates communication among the buds on a plant. Charles Darwin and his son Francis first inferred the existence of this important hormone in plants when they observed how seedlings always grow toward the source of light, and they do so by growing more on the side of the seedling that receives less exposure to light (fig. 2.8). Although bending of the seedling toward light occurred below the seedling's growing tip, the Darwins' experiment showed that light is apparently detected at the tip of the seedling. By covering the apical bud of the seedling with opaque, black caps or actually removing the buds, they demonstrated that the apical bud was necessary for the seedling to respond to light. They observed in the 1880 book *The Power of Movement in Plants* that "the uppermost part alone is sensitive to light, and transmits an influence to the lower part, causing it to bend."

Two alternative hypotheses were later proposed to account for the striking influence of light on growing tips of seedlings. Exposure of seedlings to light could result either in movement of some growth-stimulating influence to the shaded side of the seedling's growing tip or in the destruction of this influence on the side exposed to light. The former hypothesis turned out to be the correct one.

Every summer day, the growing apical buds of sunflowers respond to light. The leaves and apical buds of sunflower seedlings track the movement of the sun across the sky, facing east toward the morning

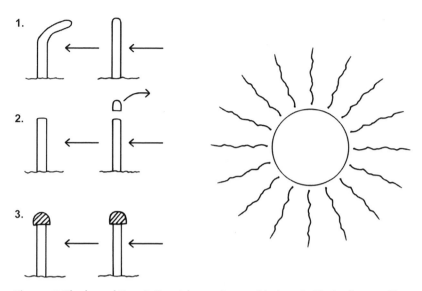

Figure 2.8 Charles and Francis Darwin's experiment with the apical buds of oat seedlings presented evidence for the existence of a growth-promoting factor or hormone. OBSERVE: (1) The oat seedling bends toward light coming from the right. The oat seedling does not bend toward light if its apical bud is (2) removed or (3) covered by an opaque cap. HYPOTHESIS: The apical bud of the seedling detects light. Some growth-promoting factor is transmitted to the lower part of the seedling that stimulates more growth on the seedling's side away from light.

sun, facing west toward the setting sun. The side of the seedling away from the sun grows more than the side of the sunflower facing the sun. The influence of sunlight shifts the growth-promoting hormone from the west side of the sunflower stem in the morning to the east side of the stem in the afternoon.

More than a hundred years after the Darwins' first experiments with the apical buds of seedlings, we now know not only the chemical structure of this simple messenger that is transported through the plant in response to light but also some of the diverse ways in which it influences the lives of plant cells. On the shaded side of the plant, the chemical messenger stimulates the growth of cells. But this messenger can influence growth in different ways that depend on its location within the plant—whether it is in the root or in the shoot. As we discovered in chapter 1, when a germinating seed or a potted plant is moved from a vertical position to a horizontal position, this chemical messenger stimulates growth on the upper side of the root but not the lower side of the shoot. The root now grows down, while the shoot now grows up. This one simple chemical instigates a whole cascade of events that ultimately result not only in shaping specific parts of plants such as leaves, fruits, and roots, but also in shaping the whole plant. This chemical message is known as auxin (*auxe* = to grow). We now know that auxin acts in concert with other plant hormones to control so many other pivotal events in the lives of plants such as aging, germination, growth, movements, and defensive responses to insects and microbes. The fate of undifferentiated plant cells—whether they become roots or shoots or both—can be determined by the simple interplay of auxin with one or more of these other hormones (fig. 2.9).

Many hypotheses have been tested and remain to be tested as plant biologists seek to understand more about the complex interplay among auxin and other hormones known as cytokinins, gibberellic acid, abscisic acid, ethylene, salicylic acid, and jasmonic acid (appendix A, fig. 9.4). The interactions among these plant hormones are complex and still being unraveled by scientists; but a few basic facts about these hormones always seem to hold true. Cytokinins promote cell divisions, while gibberellic acid promotes elongation of cells. Ethylene can inhibit the growth-promoting action of auxin and cytokinins during aging and senescence of leaves, while abscisic acid inhibits the growth-promoting action of gibberellic acid, auxin, and cytokinins during seed germination. These hormones act together—sometimes promoting, sometimes inhibiting one another's actions. Salicylic acid and jasmonic acid are produced in response to attacks from insects or fungi. And there are undoubtedly more hormones yet to be discovered, such as those recently discovered hormones known as brassinosteroids (*brassica* = cabbage; *ino* = belonging to; *steroid* = hormone) that affect plant-cell size and bear a striking resemblance to our own steroid sex hormones.

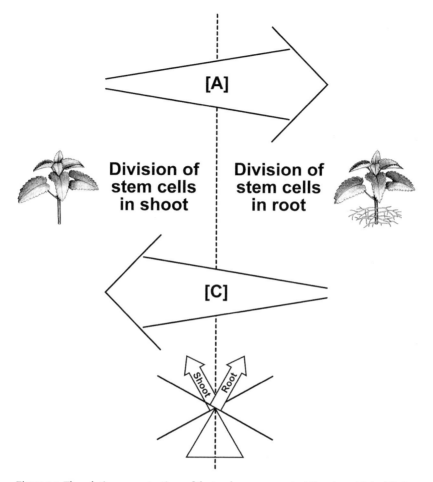

Figure 2.9 The relative concentrations of the two hormones auxin (A) and cytokinin (C) rise and fall within shoot and root meristems, orchestrating the division and differentiation of stem cells. The balance of these two hormone levels determines the fate of shoots and roots. Elevating the auxin level in an apical shoot and increasing the auxin:cytokinin ratio prompts the formation of adventitious roots on the shoot.

Auxin is a simple organic compound and is readily synthesized in a laboratory. Having ready access to auxin enables us to test hypotheses about what happens to plant tissues when synthetic auxin is used to replace a natural source of auxin or how cells and tissues respond when they are exposed to excess auxin. Auxin is usually sold as a rooting powder that is applied to aboveground cuttings of garden plants to encourage their formation of roots. This addition increases the plant's ratio of auxin to cytokinins; this change in the balance of hormones prompts the formation of roots on such cuttings, as illustrated by the diagram in figure 2.9.

OBSERVE: Dominance of a bud is assigned according to its rank along the main stem, with the highest rank at the top of the stem. If the topmost growing bud of a *Coleus* plant, a bean plant, or a basil plant is removed—this natural source of the hormone auxin—what happens to the lateral buds lower down the stem? Development of lateral buds and their stem cells along the length of a stem is apparently inhibited by the presence of cells in the topmost (apical) bud of the stem.

Start with four similar bean plants that have recently sprouted. Each plant should have two leaves opposite each other and a growing apical shoot halfway between them. Where the apical shoot joins the two opposite leaves, two lateral buds form in the angles between the leaf petioles and the vertical shoot. The angles are called axils, and the lateral buds are referred to as axillary buds. Compare what happens to these dormant buds (1) when the vertical shoot and its leaves are removed and (2) when the shoot is left untouched (fig. 2.10). Remove the vertical shoot and its apical bud from the two remaining bean plants. The apical shoot is known to be a source of auxin that exerts its apical dominance over buds lower on the stem. Could addition of auxin act as a good substitute for the apical bud that has been removed? Treat the cut surface of one shoot with an auxin paste that can be purchased at garden supply stores, but apply a paste made from flour to the cut surface of the second bean shoot as a control treatment.

Even though a potato is found underground, it actually represents a swollen stem with multiple buds ("eyes"; see chapter 3 and fig. 3.4 for more information on the lives of potatoes and other underground vegetables). Planting a whole potato results in the

Figure 2.10 Bean seedlings have a central apical shoot flanked by the two first leaves. The apical shoot is the main source for the growth hormone auxin. Apical dominance exerted by the auxin of this shoot influences the growth and form of the entire plant. If this source is removed, observe the changes that soon begin to occur in the growth of plant parts such as the lateral meristematic buds located below the apical shoot (*arrowheads*) and how this growth influences the form of the bean plant. Can the apical dominance exerted by auxin produced in the apical shoot be replaced by simply applying a paste of auxin powder?

sprouting of a single potato shoot; a single apical bud of the potato therefore expresses its dominance over the growth of the other buds. However, if the potato is cut into several pieces, with each piece having an eye, or bud, the individual buds of the potato are released from the dominating influence of a single apical bud. When

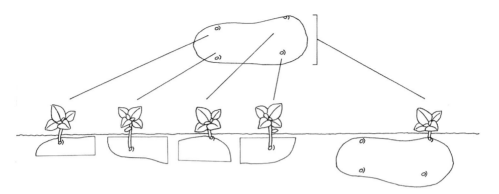

Figure 2.11 How the "eyes" of a potato determine its destiny. When the potato is cut into quarters and each is planted (*left*), the bud on each quarter gives rise to a new plant; however, if the whole potato is planted (*right*), one bud exerts dominance over the other three buds.

each piece of the divided potato is planted, each will sprout a single shoot (fig. 2.11).

HYPOTHESIZE: What do you suppose is happening in these buds, or meristematic regions, now that the apical meristem has been removed that was not happening while the apical meristem was still present? What do you think will happen if all buds between the top bud and the bottom lateral bud—the two buds most distant from each other—are removed? In the absence of intermediate buds, can the top bud still suppress growth of the bottom bud? Notice how sizes of lower, lateral buds change with their distance from the apical bud. How does the distance of a lateral bud from the apical meristem influence its growth and size?

Brussels sprouts are vegetables that are harvested in the fall when many lateral buds, or sprouts, have formed all along the length of their long stems (fig. 2.12). Those who grow this vegetable know that at the end of its growing season the best way to produce the largest sprouts for autumn harvest is to remove the large topmost bud. The removal of this one dominant bud removes the inhibition that kept

Figure 2.12 The lateral sprouts of a Brussels sprouts plant are prepared for harvest by removing the apical growing tip of the plant when these sprouts (buds) are about a half inch in diameter. With the removal of apical dominance, in about a month these sprouts will all grow to be about uniform in size and ready for harvest.

growth of the lower lateral buds in check. As you continue to harvest sprouts from the top of the plant, those sprouts lower down on the stem are released from the inhibitory influence of their more apical neighbors. As these sprouts grow, notice the precisely and orderly way they arrange themselves in spirals along the length of the stem.

Many of these beautiful forms and patterns in nature that we so often describe in appreciative, aesthetic terms can also be described in physical and mathematical terms. The Scottish biologist D'Arcy

Wentworth Thompson shared this view of the mathematical beauty of nature throughout his 1917 book *On Growth and Form*: "Cell and tissue, shell and bone, leaf and flower, are so many portions of matter, and it is in obedience to the laws of physics that their particles have been moved, molded, and conformed."

The Geometry of Plant Growth

What determines how the sprouts, buds, and first leaves are assigned their positions around the topmost apical bud of a newly sprouted plant? Peer down the length of a stem from the apical meristem and you will see how sunlight is evenly distributed to surfaces of the leaves below. Each leaf receives its fair share of sunshine by having leaves arranged in orderly spirals around stems. This spiral arrangement of leaves around stems assures that no leaves are completely shaded by leaves above them (fig. 2.13). The seeds of sunflowers, the berries of pineapples, and the scales of pinecones and acorn caps are also spirally arranged so that the maximum number of seeds and scales and berries can be packed into the space provided. These orderly arrangements have been mathematically described by a series of numbers beginning with 0 and 1. The remainder of the Fibonacci series is derived by repeatedly adding the two consecutive numbers to obtain the next number in the series—0, 1, 1, 2, 3, 5, 8, 13, 21, 34, 55, 89, 144, 233 . . .

OBSERVE: Start at the topmost leaf on a stem and assign it the number 0. Then count the number of leaves that are spirally arranged along the stem until you turn 360 degrees and reach the leaf that occupies space below leaf 0. Continue counting leaves along the stem until two more complete revolutions around the stem bring you to the second and third leaves directly below the topmost leaf. Note how the differences in numbers assigned to leaves that align with each other represent numbers (e.g., 5, 8, 13) in the Fibonacci series.

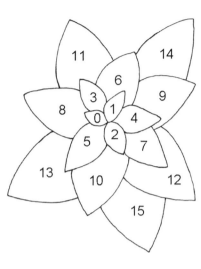

Figure 2.13 Looking down on the apical meristems of plants shows that the sizes and ages of leaves are arranged from top to bottom in an orderly spiral around the main stem of the plant. On the left is a photograph of the apical meristem of ornamental tobacco, *Nicotiana*, as viewed from directly overhead. On the right is a sketch showing the view of a sunflower's apical meristem from above and before flowers have formed.

This same series also describes the spiral patterns of seeds, flowers, and fruits. Once you start looking for these numbers and patterns, you will begin seeing them in many plants—such as sunflowers—in their arrangements of leaves, flowers, and seeds.

Take, for example, the distinctive scales on the cones of evergreen trees. Unlike the flowering plants of the garden in which seeds are encased in fruits, the seeds of evergreen trees are exposed between the scales of their cones. These scales are arrayed in interlacing spiral patterns; the numbers of spirals along the length and at the base of the cones correspond to numbers in the Fibonacci series

Figure 2.14 In the cones of spruce, pine, fir, and hemlock (*top, left to right*) spiral patterns intertwine; the numbers of their turns correspond to numbers in the Fibonacci series. Spirals are easy to count at the base of each pine cone shown in the bottom row. The one shown here has eight and thirteen spirals; others, such as the cone of white pine (*top row, second cone*), have five spirals.

(fig. 2.14). On the cones of fir trees (fig. 2.14, top row, third cone), look for the spiny scales that have been described as the "tail ends of mice" ducking under regular smooth scales. These spiny scales stand out so well from the other scales that you can easily distinguish their spiral patterns from the surrounding spiral patterns of the smooth scales. The spiral patterns displayed by the evergreen cones of gymnosperms are perfectly mirrored in the angiosperm fruits of pineapples and the flower buds of artichokes.

HYPOTHESIZE: At any given height above the soil surface, the number of branches on a bushy garden plant such as basil, catnip, or *Coleus* is a number in the Fibonacci series (fig. 2.15). What happens to the orderliness of this Fibonacci series if we perturb the three-dimensional arrangement of branches on a plant? The essentially two-dimensional branching patterns of common prostrate weeds in the garden illustrate this orderly arrangement of a plant's branches even better. The number of growing tips on a major branch of one of the prostrate, horizontally spreading weeds such as common knotweed (*Polygonum*), bedstraw (*Galium*), purslane (*Portulaca*), carpetweed (*Mollugo*), or spurge (*Euphorbia*) is also a number in the Fibonacci series (fig. 2.16). Remove one of the two or three lowermost

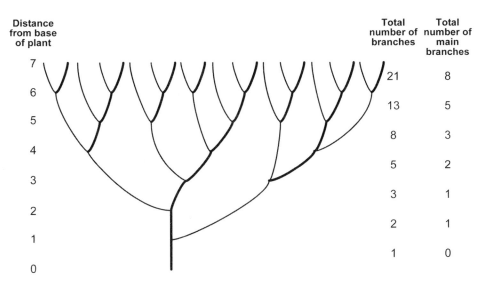

Figure 2.15 Garden plants, bushes, and trees have typical branching patterns. Each main branch (*thick line*) forks at regular, periodic levels (1–7) aboveground; each secondary branch (*thin line*) forks at every other level.

Figure 2.16 Common prostrate weeds such as spurge (*Euphorbia*, *top*) and carpetweed (*Mollugo*, *bottom*) spread across the ground in the garden, repeatedly branching in an orderly sequence expressed mathematically by the Fibonacci series.

branches of one of these rapidly growing weeds and observe how the arrangement of branches subsequently shifts over a period of several weeks. What happens if you remove one or two of the uppermost branches from the crown of one of the bushy garden plants? How do the other branches adjust to this absence of their fellow branches?

Are the positions of branches fixed around the crown of the plant in a rigid, mosaic pattern that does not change with time, or can branch positions shift and regulate to compensate for the absence of branches that have been removed from their rank?

The Orderly Birth and Death of Leaves

As a plant ages at the end of its growing season, what controls the massive, annual fall of all the leaves that takes place each autumn when petioles of leaves detach from their stems? This detachment, or abscission (*absciss* = cut off), of aging leaves is responsible for the thousands of pounds of leaves that fall from the trees on each acre of forest every autumn. Do factors other than old age alone control the dropping of leaves from a plant?

OBSERVE: An important observation about this annual death of leaves is that simply removing the blade of a leaf triggers the premature detachment of its leaf petiole. If the blade of a summer leaf of spinach or *Coleus* is removed but its petiole is left untouched, the petiole will soon lose its green color and turn prematurely yellow (fig. 2.17). Within days, a sharp transition from yellow petiole to green stem is evident; the summer petiole now falls to the ground and behaves as an old, autumn petiole would during a windy day in October or November. What hypothesis can account for this accelerated aging of a leaf?

While the plant hormones auxin, cytokinins, and gibberellic acid promote growth and slow aging of plant parts—including leaves—the hormone ethylene has been shown to counteract the actions of these three hormones in leaves. Ethylene is not only a plant hormone, but it is also a common component of gas that was once used to heat greenhouses. About a hundred years ago, growers noticed that traces of this gas in their greenhouse air caused leaves to fall prematurely from plants; the older the plant, the more its leaves were sensitive to something in the gas. Chemical analysis of the gas showed how that

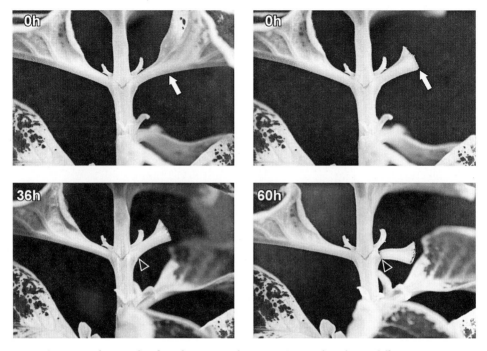

Figure 2.17 Photographs of a *Coleus* stem and two opposite petioles taken at different times (0, 36, and 60 hours) after the leaf blade belonging to the right petiole was removed (*at arrow*). The left leaf and its petiole were untouched. The zone of abscission, or detachment, of the right petiole from the main stem (*arrowhead*) developed within a few hours after its leaf blade was removed.

something turned out to be ethylene. This initial observation initiated a whole series of experiments that tested hypotheses of how ethylene influences the lives of plants and how ethylene exerts its influence through interactions with its fellow plant hormones.

HYPOTHESIZE: What is the fate of a *Coleus* petiole if three-fourths or only half of its leaf blade is removed? What can you infer about the interaction of hormones such as auxin and cytokinins that promote the growth of leaves with the hormone ethylene, which promotes aging of leaves?

What happens if you apply an auxin paste to the petiole as soon

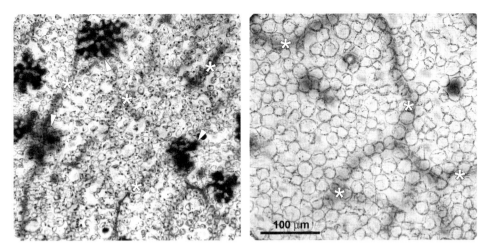

Figure 2.18 Photographs of spinach leaves show magnified views of leaf cells as they progress from young and green (*left*) to old and yellow (*right*). As the cells of the leaf age, they swell, increasing in size, and lose most of their patches of chlorophyll. The dark, winding channels (*) are the vascular bundles of xylem and phloem that lie beneath the leaves' surfaces. The five, even darker, patches (*arrowheads*) of the young spinach leaf are tannin pigment in the surface layer of cells.

as the leaf blade is removed? What happens if you wait a day before applying the auxin paste to the petiole?

The leaves of a plant have shortened lifetimes once apical flowers form on the plant. If the flowers of *Coleus* or a spinach plant are removed as soon as they begin to form, the lives of the nearby lower leaves are spared. The sooner the incipient flowers are removed, the longer the lower leaves of that plant survive and remain green. By continually harvesting the leaves of spinach and preventing its flowers from forming, the lives of spinach plants and the harvest of their leaves can be prolonged for weeks (fig. 2.18)

This removal of the apical flower buds, or "topping," is an old method used by tobacco growers who always try to maximize the size of the leaves they harvest by prolonging the leaves' lifespans and growth. Topping removes the plant's apical dominance and a main source of the hormone auxin, which inhibits growth in parts of the

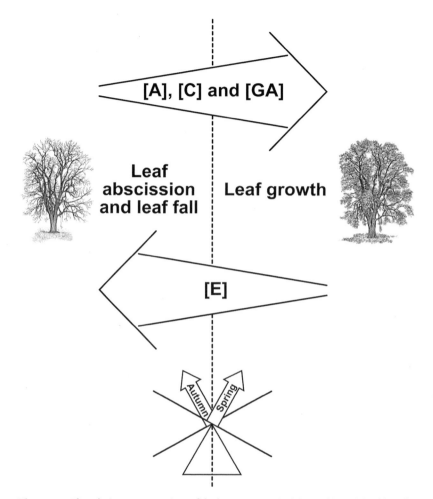

Figure 2.19 The relative concentrations of the hormones auxin (A), cytokinin (C), gibberellic acid (GA), and ethylene (E) rise and fall within leaves, orchestrating their growth, eventual aging, and abscission. The balance of hormone levels determines the fate of leaves each fall.

plant lower on the stem. When this apical source of inhibition is removed, growth increases lower down on the plant, and the leaves below the point of topping undergo a growth spurt. How the loss of auxin from the top of the plant stimulates the growth of plant parts lower down is still not understood in detail; however, recent experiments hint that in the absence of apical auxin, a substance produced even farther down in the roots is transported to the lower buds and acts to promote their growth through the action of auxin, cytokinins, and gibberellic acid.

OBSERVE: Try another experiment to see what effect topping has on the aging of older, lower leaves of a bean or tobacco plant that are starting to turn light green and yellow with age. Mark each old, yellowing leaf with a dot of liquid paper. See what happens to the aging leaves below when the younger leaves and buds of the plant above are removed. Somehow removal of this apical region not only promotes growth of lateral buds but also rejuvenates nearby aging leaves.

If the fruit pods of okra are always picked before they become hard and woody, the okra plant will continue flowering until frost comes; but if the pods of a plant are left to harden, the whole plant will soon stop flowering and begin dropping its leaves.

By studying the responses of whole plants to removal of specific plant parts, scientists have been able to piece together the puzzle of how plants and their hormones integrate the development of all their parts. As is the case with other decisions in the lives of plants, the interplay of different hormones decides what fate awaits the cells of plants (fig. 2.19). Observing the responses of plants to different experimental treatments enables us to propose hypotheses that can be further tested with new experiments.

3

THE UNDERGROUND WORLD: BULBS, TUBERS, AND ROOTS

The roots of a plant—the descendants of its seed's hypocotyl—extend countless branches into the soil in search of water and nutrients. Roots exude substances into the surrounding soil to attract and nourish the countless soil microbes that associate with roots and help make many soil nutrients available for uptake. Water and nutrients are channeled upward to nourish parts of the plant aboveground, but a share of the nutrients from the soil along with nutrients produced aboveground are also stored belowground in roots, bulbs, and tubers. These underground forms of plants represent the portion of a garden's edible harvest that lies hidden from view.

The leaves of potatoes, carrots, turnips, rutabagas, onions, and garlic channel much of the energy of the sun into

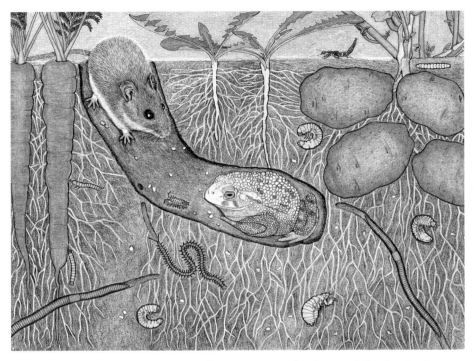

Figure 3.1 A mouse, a toad, and a woodlouse meet in a cool underground chamber. Earthworms tunnel through the nearby soil, creating passageways for the countless roots and other creatures that share the soil world. A long, sinuous soil centipede hunts along the tunnels left by earthworms, while a rove beetle with raised abdomen (*upper right*) stalks insects at the soil surface. Also in the upper right, the larva of a robber fly hunts soil insects just beneath the soil surface. Grubs of different scarab beetles and the nymph of a cicada feed on the abundance of roots. Two larvae of a click beetle munch the carrot.

forming the roots, bulbs, and tubers that we eat to obtain the energy and nutrients they have stored. What makes one garden crop a tuber, another a bulb, and a third crop a root depends on the part of the plant where most of its energy is stored and the location of buds to which this energy can be channeled when a new plant begins to form. Even though they all reside underground, tubers and bulbs are not true roots. Bulbs are short stems surrounded by underground leaves, and most tubers are underground stems that masquerade as roots but that can sprout true roots of their own under the right conditions.

The Phenomenal Speed of Growing Roots

Roots grow at phenomenal rates. A scientist once decided to see just how far and how fast the roots of one rye seed would grow. Four months after planting the seed, he washed all the soil off the roots and began the daunting task of the counting number of roots and the lengths of each. What he found was that in this relatively short period of time, this one rye plant had extended 15 million roots, and that the total length of these roots was 380 miles. However, if he also counted all the tiny root hairs (fig. 1.7) covering these innumerable roots, the length of the roots now shot up to seven thousand miles.

OBSERVE: We can clock the speed at which the first root of a seed of rye grass travels in a 100-millimeter petri dish. Place a seed in the center of a dish whose bottom is covered with moist filter paper; its destination will be the edge of the dish. Within a day, the first roots will sprout from the seed and begin their journeys (fig. 3.2). Which rye seed has the fastest-growing and straightest-growing root that enables it to reach its destination first?

HYPOTHESIZE: Hidden as they are in darkness and tiny spaces, we know very little about the underground lives of roots. Some scientists now have evidence not only that plants can distinguish roots of their own species from roots of other plant species but they can

Figure 3.2 Roots and their root hairs rapidly grow from two rye seeds over a two-day period (24 to 72 hours).

actively inhibit growth of roots that they somehow recognize as be-longing to different plant species. Set up three 100-millimeter petri dishes with moist filter paper. Place a rye seed in the center of two of the dishes and three rye seeds together in the center of the third dish. Add two seeds of Kentucky bluegrass adjacent to the rye seed in one dish; add two seeds of barley adjacent to the rye seed in the second dish. Do roots of different plant species come in contact with each other? How is root growth of a species affected by (1) the pres-ence of another species and (2) the presence of another member of its own species?

The Ability of Bulbs, Tubers, and Roots to Form New Plants—No Seeds Required

Bulbs, tubers, and roots—all are often referred to as root vegetables, even though the only true roots among them are carrots, parsnips, beets, turnips, rutabagas, and radishes. All root vegetables, however, are underground organs that store excess sugars produced by photo-synthesis. Potatoes and sweet potatoes store their excess sugars mainly in the form of starch. Starch is a chain, or polymer, of many sugars strung together. Whenever the whole plant needs energy, it can tap this underground reserve by breaking down the long chains of starch into smaller sugar units and transporting them to energy-demanding buds. The true root vegetables—along with bulbs that are members of the onion family (garlic, onions, and leeks)—store a little starch, but mostly sugars, at the end of summer.

Most roots, bulbs, and tubers are organs of biennial or peren-nial plants that store sugars in their roots not only during their first winter but also during subsequent winters. Whatever sugars and starches are stored underground at the end of each growing season will supply the energy for growth and flowering aboveground the fol-lowing year.

In autumn, phloem channels of plants transport starch and sug-ars to be stored underground for the winter. In early spring, as the buds of plants rapidly expand into developing leaves and flowers, these reserves of energy that have been stored in the parenchyma (*par* = beside; *enchyma* = to insert) cells of roots, bulbs, and tubers are transported via the vascular transporting system to the energy-demanding buds aboveground. Normally the division of labor between the xylem and phloem cells of the plant's vascular system is clear-cut. The long, hollow xylem cells act as conduits for water and mineral nutrients from the soil, and the phloem channels transport those sugars that form wherever photosynthesis is happening. But in the spring, xylem cells also serve as sugar transporters. Think of the sap that is collected by the gallon each spring from sugar maple trees. This sweet sap is rising from the roots through the tree's xylem channels. Likewise in late summer and autumn, phloem cells can transport much-needed water and nutrients to developing fruits and seeds. When special demands are made on the vascular transport system of a plant, an exchange of cargo is now known to occur at these times between xylem channels and phloem channels via special cells, called transfer cells, that connect these two channels.

OBSERVE: A special solution (I-KI) containing iodine (I) and potassium iodide (KI) has been used as a specific stain for starch in plant cells. This solution is referred to as Lugol's solution and is available from biological supply houses. Cut a thin slice of plant tissue with a sharp knife or a single-edge razor blade. To the freshly cut tissue add enough drops of I-KI to cover the area of tissue that you are examining. Any starch that is present will appear as discrete granules within individual cells and will stain brown or blue-black (fig. 3.3). These plant-tissue slices can be rinsed with tap water and placed on microscope slides for closer inspection of individual cells. Placing a thin cover of glass over the stained tissue slice will sandwich the plant cells between two layers of glass for best viewing. Whatever starch granules are present are stored in chloroplasts that are specialized for starch storage, called amyloplasts (*amylo* = starch; *plast* = form). When sugar is needed somewhere in the plant, these

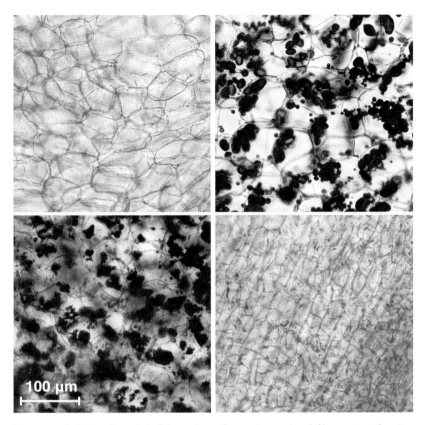

Figure 3.3 Staining of starch in thin sections of a turnip root (*top left*), a potato tuber (*top right*), a sweet potato tuber (*bottom left*), and a carrot root (*bottom right*). Parenchyma cells of these plants are specialized for the storage of starches and sugars.

amyloplasts convert their starch granules back into sugar that can be transported to developing buds and meristems, where they are most needed.

How do different roots, bulbs, and tubers compare in the intensity of their staining with I-KI? Which vegetables contain the most starch, and which vegetables are the sweetest?

Some roots, bulbs, and tubers improve their flavor while some lose their flavor during storage. As seeds begin germinating and as roots, bulbs, and tubers begin sprouting, nutrient reserves stored in their cells are mobilized to power their new spurts of growth. What are the optimal conditions for maintaining the best flavors of underground vegetables or for even enhancing their flavors? Does their change in taste reflect a loss or gain of starch granules and amyloplasts from cells as the balance of starch and sugar shifts with time in storage?

For generations many gardeners have stored their harvest of roots, bulbs, and tubers through the months of winter in dark, well-insulated root cellars. From many years of experience, farmers and gardeners have discovered the best conditions to use for long-term storage of each type of vegetable. Leaving only about an inch or half inch of top on the root vegetables (carrots, parsnips, beets, turnips, rutabagas) minimizes the ongoing transpiration of water from the leaves and helps maintain their crispness during storage. Roots are stored best at cold but not freezing temperatures (33 to 40 degrees Fahrenheit = 1 to 4 degrees Celsius). While these vegetables store best at 33 degrees Fahrenheit, temperatures colder than 38 degrees Fahrenheit (3 degrees Celsius) cause potatoes to lose their flavor. Sweet potatoes store best at temperatures between 55 and 60 degrees Fahrenheit (12 to 16 degrees Celsius). Onions are best stored between 40 and 50 degrees Fahrenheit (4 and 10 degrees Celsius). Potatoes and onions, however, should not be stored together; potatoes take on the odor of onions, and onions spoil by taking on moisture from potatoes. Placing a ripening apple in the storage area with potatoes and sweet potatoes will discourage the latter from sprouting. What hormone do ripe fruits emit that could influence sprouting of tubers? See figure 4.15. But if carrots are nearby, this hormone emitted from ripening apples will induce the formation of a compound in the carrot that confers a bitter flavor. The experiments and experiences of earlier generations of gardeners have established these optimal conditions for maintaining the best flavors of underground crops stored through the months of winter.

HYPOTHESIZE: True root vegetables—carrots, beets, parsnips, turnips, and rutabagas—are all started from seeds. Potatoes, sweet potatoes, garlic, and onions have seeds also, but rarely does anyone start these tubers and bulbs from seeds. For one thing, plants of these latter four vegetables that are started from seeds take far longer to develop than those started from shoots, mini-tubers, or mini-bulbs. Garlic and onions are usually started from cloves or mini-bulbs, while potatoes and sweet potatoes are started from mini-tubers or green shoots. New plants of garlic, onions, potatoes, and sweet potatoes all sprout from meristematic buds—often numerous—nestled in their bulbs and tubers. Each true root vegetable, however, that sprouts far more readily from a seed than from a bud, has only one apical bud from which a new plant can sprout.

Potato tubers are swollen stems that grow underground; and like stems aboveground, these underground stems have many buds scattered over their surfaces. If exposed to light, potato tubers will become green like stems aboveground. Each potato tuber has breathing pores called lenticels—the same breathing pores that are found aboveground on stems. All these features of potato tubers mark them as stems that inhabit the underground and store countless starch granules.

Cut a potato along its long axis, and then cut another potato along its transverse axis (fig. 3.4). Within the starchy, white background, you will see a core of clearer tissue that runs along the long axis of each potato. This represents the central portion or vascular system of the underground stem. This central channel sends a branch of vascular conducting tissue to each bud of the potato, just as each bud found on plants aboveground is connected to conducting channels of the plant's main stem.

Onions, garlic, and leeks are neither tubers nor roots. The energy-rich parts of these plants that we eat are called bulbs (fig. 3.5). Most of each onion bulb is made up of unusual, colorless leaves called scale leaves that are arranged in circles around a central stem and its apical bud. It is an underground bud surrounded by its protective scale leaves. In the case of a garlic bulb, several lateral buds (cloves) sprout from the central stem. The special scale leaves that are arranged in

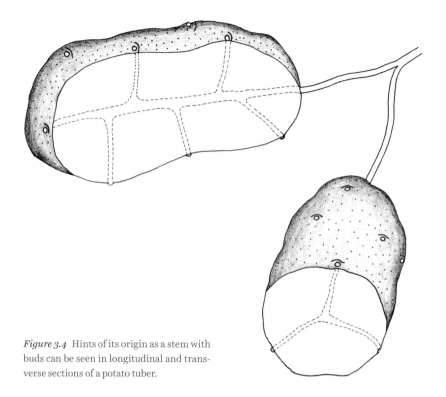

Figure 3.4 Hints of its origin as a stem with buds can be seen in longitudinal and transverse sections of a potato tuber.

rings around the central bud of an onion or leek bulb behave like leaves of other plants that are concentrically arranged around a single plate-shaped stem. The base of each small stem sprouts a cluster of roots. Garlic, however, represents a mother bulb that has given birth to several daughter bulbs, each daughter bulb referred to as a clove; and each daughter bulb has a stem that can sprout from its base and later give rise to granddaughter bulbs.

Roots, bulbs, and tubers have the special ability to generate whole, new plants without germinating a single seed. Special portions of each root, tuber, or bulb are made up of stem cells, which are totipotent (*toti* = all; *potent* = powerful); that is, they have the ability to form all parts of a plant—aboveground and belowground—even though they start out as only the underground parts of plants. How do tubers, roots, and bulbs differ in their placement of these totipotent

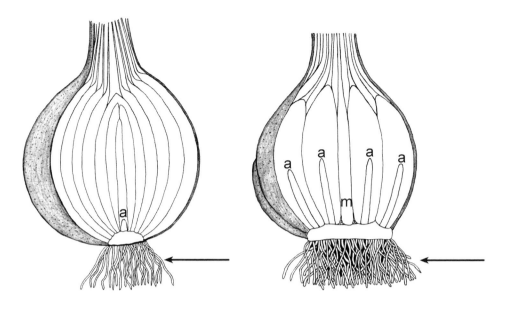

Onion Bulb   Garlic Bulb

Figure 3.5 Left: Slicing through the middle of an onion bulb reveals the chlorophyll-free leaves (scale leaves) that are arranged concentrically around the short central stem with its apical bud (a) and its basal roots (*arrow*). *Right:* A garlic bulb represents a cluster of garlic cloves, all descended as lateral bulbs from an initial, centrally located mother stem (m). Each daughter bulb has its own central stem with apical bud (a) and basal roots (*arrow*).

cells? What is the smallest portion of a tuber, a bulb, or a root that can give rise to a new plant?

Transporting Water and Nutrients along Root Channels

Living root cells of plants, like all living cells, contain relatively high concentrations of many different chemicals, including simple mineral nutrients, vitamins, sugars, proteins, and nucleic acids dissolved in relatively small volumes of water. As they take up nutrients from the soil, membranes of root cells use energy to pump mineral nutrients from outside the cell to the interior of the cell. In the adjacent soil, mineral nutrients exist in far lower concentrations dissolved in relatively large volumes of water. Water moves from its higher concentration in the soil across the membranes of the root cells into regions with lower concentrations of water. Water thus "pushes" its way by osmosis (*osmos* = push) as it moves from the soil into root cells (fig. 3.6). Although the living root cells allow water to pass freely, their membranes selectively prevent the outward movement of all other chemical substances. The plant cell subsequently swells from its uptake of water until the water or turgor (*turgo* = swollen) pressure squeezes water to adjacent cells. Normal turgor pressure in plant cells imparts a crispness and firmness to plant tissues; it stretches the walls of the cells and helps cells grow in size. A loss in turgor pressure of cells in a tissue—when water no longer exerts pressure against the cells' walls—results in tissues that are limp, wilted, and shriveled. When wilted plant cells are placed in water again, the water pushes its way into the water-deprived cells by osmosis and stretches each cell from wall to wall.

OBSERVE: The "pushing" of water from its higher concentration in soil belowground to its lower concentration in the leaves aboveground—root pressure—can be graphically demonstrated with a peeled potato tuber placed in pure water, some sugary syrup placed in a hollowed-out chamber in the core of the potato, a tightly fitting one-hole stopper placed at the entrance to the chamber, and a narrow glass tube placed in the center of the stopper (fig. 3.7). Peeling the skin of the potato removes those outermost cells that impede the flow of water into the tuber. The glass tube models the water-conducting xylem tubes, and the potato in the beaker of water models a plant root surrounded by soil water. From its higher concentration in the beaker, water progressively pushes its way by osmosis to its lower concentration in the peripheral cells of the potato tuber before pushing to an even lower concentration of water in the center of the tuber. Here water has been displaced and its concentration has

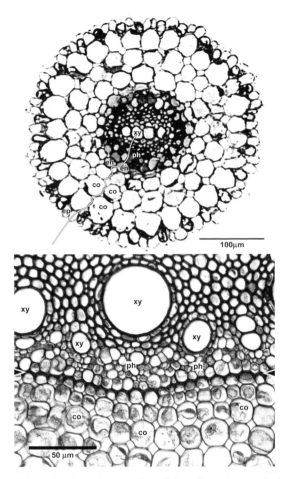

Figure 3.6 Top: This section through a root of a radish seedling gives us a look at the cellular landscape that water must cross as it moves by osmosis from soil to the water-conducting xylem vessels (xy) in the center of the root. The long, sinuous arrow marks one of the countless paths that water follows on its way. Some water also passes from one cell to another via microscopic channels that connect cells. Water pushes its way by osmosis through the cells of the outermost epidermis (ep) and flows among and around the cells of the cortex (co) until it reaches the ring of cells known as the endodermis (*endo* = interior; *dermis* = skin). *Bottom:* The cell walls of the endodermis layer (*arrowheads*) are waxy and thick, acting as barriers to the flow of water and minerals around and between cells. Endodermis is the gatekeeper to the root's vascular system, demanding that transport be through them and not around them. When water from the soil reaches this layer of cells after first moving around and through cells of the cortex (co), osmosis again drives it across the cells of the endodermis, then special membrane proteins selectively transport or exclude specific mineral nutrients. After crossing the endodermis, water flows around phloem cells (ph) on its way to enter the hollow and perforated conducting cells of xylem (xy) in the center of the root.

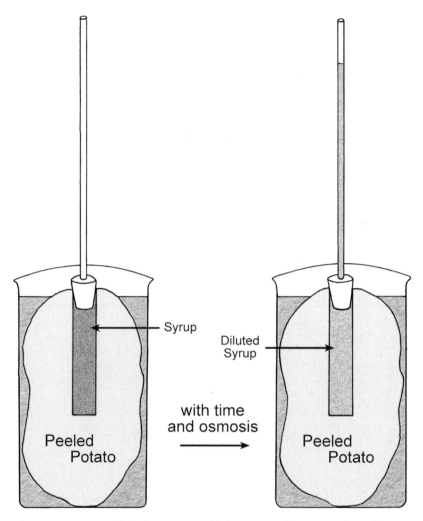

Figure 3.7 A potato model that is easy to assemble demonstrates how osmosis drives the movement of water from the soil, across root cells, and up xylem channels to leaves aboveground.

been reduced by a high concentration of sugars and minerals in the "syrupy" center of the tuber. The central core of the potato has been replaced with corn syrup to simulate a high concentration of sugars and minerals that are normally found in the vascular conducting cells of phloem and xylem in the center of roots. The "push" of the water is reflected in the rise of the liquid column in the glass tube. How rapidly does the water push its way up the glass tube? When does the water stop rising in the tube? Eventually the pressure exerted by the water that has risen in the glass tube begins to push water out of the potato as rapidly as water enters the potato.

OBSERVE: This buildup of osmotic pressure in the roots of plants is revealed on mornings after cool, humid nights by droplets of water that appear along the edges of leaves. After dark on a humid night, this root pressure, or osmotic pressure, builds up to a point where sap is forced out on leaf surfaces in liquid form, not the vapor form of daytime transpiration (*trans* = across; *spiro* = breathe) or evaporation from leaf surfaces (chapter 6). This sap contains not just water but also any minerals or chemicals—even pesticides—that roots have taken up from the soil. By evening, the pores, or stomata, of leaves have shut; transpiration of water from these pores and its evaporation as water vapor has ceased for the day. Evaporation of water from the innumerable stomata of transpiring leaves no longer draws water upward in the xylem vessels. At night, root pressure from below moves the sap upward—not the evaporation from above.

This sap that exudes as droplets from particular points on the margins of leaves is often mistaken for morning dew (fig. 3.8). On leaves of grass seedlings such as rye, the sap forms as a pearl droplet at the tip of each leaf. Morning dew, however, does not take this form but settles uniformly over the cool surfaces of leaves as condensation from the surrounding air. Dew is pure water without any of the minerals or chemicals found in the xylem sap expelled from leaves by root pressure.

These plants are demonstrating what is known as guttation (*gutta* = drop). Under these environmental conditions, the root pressure squeezes water out of special glands found along the margins of certain plants. The glands are known as hydathodes (*hydat* = watery; *hod* = way); their locations on leaves are evident by the

Figure 3.8 Left: Strawberry plants exude xylem sap from hydathodes at the tips of their leaves on a cool, humid summer morning (Ed Reschke/Getty Images). *Right:* Each hydathode (*black arrow*) is a specialized arrangement of cells at the end of a leaf vein (*arrowheads*) through which sap is discharged from the vein to the surface of the leaf. Some of the leaf's stomata are marked with small white arrows.

arrangement of water droplets on those garden plants that display guttation on cool, humid mornings. These hydathodes act as pressure relief valves for the xylem sap. Leaves of strawberries, grapes, tomatoes, grasses, and roses are among those garden plants where the process of guttation is most conspicuous.

HYPOTHESIZE: A closer look at the process of guttation helps develop a better appreciation for how plants move water and nutrients through the xylem vessels of their vascular systems. Strawberry plants show the best example of guttation; but rye seedlings and tomato plants are also good experimental subjects that readily undergo guttation on certain summer nights. What environmental conditions can promote guttation? Rather than waiting for the right environmental conditions to occur outdoors, you can control the exposure of potted plants to a variety of soil and air conditions hypothesized to induce—or fail to induce—guttation of tomato, strawberry, or rye leaves. Since guttation reflects the root pressure exerted in xylem vessels, then guttation from leaves should increase as their root pressures increase. Whatever environmental conditions elevate root pressure should also enhance guttation.

(1) Increase the concentration of mineral nutrients for one plant by adding a teaspoon of fertilizer to the potted soil but omitting this addition of fertilizer from the other pot. Leave the two plants outdoors and check them each morning until water droplets appear at the leaf tips of rye seedlings or on the margins of tomato and strawberry leaves. What temperature conditions promote guttation? (2) Increase the soil moisture in one pot while not adding additional water to another pot. Leave this pair of pots with plants outdoors until water droplets appear one morning on the leaf edges. What special conditions induced guttation this time? (3) After two tomato plants, two strawberry plants, or two rye seedlings have spent a very hot day in the warm sun, leave one plant outdoors and place its partner in an air-conditioned room. What happens if the temperature drops more than 10 degrees Fahrenheit outdoors? What influence does the humidity of the air-conditioned room exert on the process of guttation? (4) Do you observe any differences in the guttation for smaller, younger leaves and larger, older leaves?

OBSERVE: Nutrients and water from the soil travel through specific channels on their way from the soil to destinations among the leaves, flowers, and fruits. These channels are created by a long line of hollow, thick-walled cells arranged end to end like the pipes of a pipeline extending from root tips to leaf tips. Having lost their end walls and all their contents except for their sturdy side walls, these long, cylindrical xylem cells have become hollow channels, specialized for their job of conducting sap from roots to plant parts aboveground. You can follow the passage of nutrients and water from the soil into parts of a plant high above the ground by placing the base of a celery stem in a nutrient solution that has been marked with bright food coloring. After the colored solution has traveled to leaves at the top of the celery stalk, cut the stalk and see where the color is concentrated. These represent the "pipelines" directing nutrients and water from the soil to leaves, stems, and flowers aboveground.

Different vegetable roots have their own distinctive xylem pipelines. By highlighting the xylem channels that guide water and nutrients from the soil to all parts of the plant with food coloring, we can see that the design of these channels in each root has its own specific layout. Place the tapered tips of a carrot root, a radish or turnip root, and the root of a golden beet in a solution of blue dye. After several hours rinse the color from the outer surface of each root before using a sharp knife to slice each transversely and lengthwise. The staining patterns revealed in roots by their uptake of dye tell us about the arrangement of their xylem pipelines (fig. 3.9).

Radishes and turnips are members of the same plant family—known to many as the cabbage or mustard family—so we predict that the arrangements of their water-conducting channels are similar. When the tip of a turnip or radish root is left in a dye solution for several hours, the dye is channeled along the same routes that water

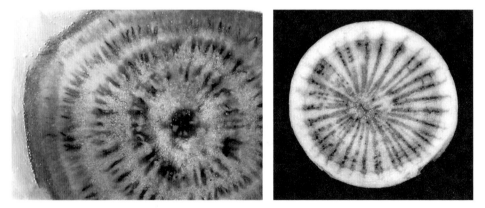

Figure 3.9 Dark food coloring travels along the concentrically arranged water-conducting channels in the root of a golden beet (*left*). Dye travels through the central xylem water-conducting channel in the turnip root (*right*) as it does in the carrot root shown in figure 2.5.

and mineral nutrients are transported from the soil to all parts of the turnip and radish plant.

The meristematic cambium cells of vegetable roots divide toward the inside and toward the outside—to produce not only more unspecialized stem cells like themselves in the cambium, but also specialized phloem cells toward the outside of the root and another class of specialized xylem cells toward the center of the root that conduct water and mineral nutrients from the soil. Each turnip, radish, and carrot root has a single ring of these cambial stem cells (fig. 2.5).

Beet roots (of the goosefoot family) do things differently. Not only are the pigments of beets and their relative Swiss chard chemically different from pigments of most other vegetables, but they also have a different arrangement of root channels for transporting water and nutrients. The concentric rings of a beet root tell us about the age of a beet—just as tree rings tell us about the age of a tree. Rings of beet roots are added on a weekly basis and represent alternating growth rings of cells. The older and larger the root, the more rings of cells. In the case of beet roots, stem cells of the cambium appear as multiple concentric rings separated by cells specialized for transport of nutrients (xylem and phloem) and the parenchyma cells specialized for storage of nutrients. Stem cells within each ring divide toward the periphery of the root to become the specialized phloem cells that conduct sugars from leaves above, while stem cells divide toward the center of the root to form the specialized xylem cells that carry water and nutrients from the soil.

OBSERVE: The transport channels that send water and nutrients upward are dead, hollow cells whose cell walls act as pipelines (fig. 3.10). The long, hollow tubes of xylem cells are arranged end to end along the length of roots and stems in parallel bundles and are connected both vertically and horizontally on their sides by small pores called pits, such that if one channel is blocked by air or an invading fungus, the water transport can be shifted to one or more new channels.

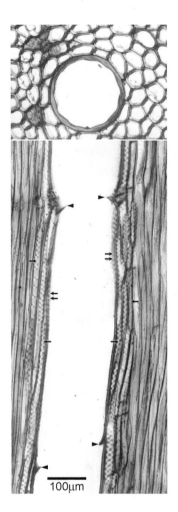

Figure 3.10 Top: In a cross-section of an oak trunk, a large hollow xylem vessel stands out in the midst of smaller, hollow tracheid cells. *Bottom:* When viewed from the side in a longitudinal section of the trunk, the large hollow vessel and smaller hollow tracheid cells are arranged end to end and side by side as parallel pipelines directing the flow of water up roots and stems. The remains of the cell walls that once separated the vessel cells of one pipeline (*arrowheads*) have disintegrated, creating a long, uninterrupted vertical channel. The thinner conducting cells, called tracheids, surrounding the large vessel cell are tapered at both ends. Numerous pits, or pores (*single arrows*), connect tracheid cells at their ends and all along their lengths. The pits of vessel cells (*double arrows*) are located on their sides. In the xylem tissue of ferns and conifers, such as pine and spruce, vessels are absent and tracheids are the only xylem conducting cells.

How does the upward movement of sap through xylem vessels on a warm summer day compare with the movement of sap from the roots to the tips of branches on a warm spring day before the first leaves have appeared? In early spring there are as yet no leaves expelling water from their innumerable stomata. However, as water evaporates at the leaf surface on a summer day, even more water is drawn upward in the xylem channels. Water in hollow xylem channels that stretch from root tips to leaf tips is drawn upward along the entire height of a plant (this massive movement and transpiration of water from leaves is discussed in chapter 6). Removal of water by evaporation at the top of the xylem channels draws water upward in the same way that drinking draws water upward along the entire length of a drinking straw.

During the growing season, as water evaporates from surfaces of leaves through the myriad stomata that open and close in concert with transpiration, the sap containing water and soil nutrients in xylem vessels is pulled upward from the roots. Also during the growing season, the sugars produced by photosynthesis in leaves are carried around the plant in the phloem cells. In the autumn, these sugars are channeled downward and stored in the roots as starch. In the spring, however, when the starch stored underground is converted to sugar, the transport channels in plant stems must move sugars upward; and only at this time in the annual affairs of trees does the upward movement of sugar and sap take place in the xylem vessels. Sap from these xylem vessels of maple trees is tapped each spring for making maple syrup.

The sap from maple trees flows freely on warm days that follow freezing nights. Not all plants, however, have free-flowing sap that can be tapped in the spring. Saps from other trees such as willows, elms, ashes, and oaks do not flow on warm days in the early spring. Scientists have hypothesized that differences in the structures of their xylem channels account for why we tap maple trees for their sap rather than other trees. In maple trees, the xylem vessels are filled with sap that cools after dark; the gases within the vessels (such as carbon dioxide) now dissolve in the cooler sap; their pressures drop, and the increase in sugar and gas concentration draws water by osmosis from a series of surrounding cells leading all the way back to the root tips. As night temperatures continue to drop, water freezes inside the hollow vessels and the dissolved gases are further compressed. Not until the sun warms the vessels the next morning, thawing their sap, do their gases expel from the sap and expand, increasing the pressure within the xylem channels and forcing the sap upward again.

A cycle of freezing and thawing is essential to drive the flow of sap in maple trees on spring days before their leaves have formed. Before buds expand and leaves begin unfolding in the early spring, prune the ends of different tree branches, including a maple branch. Observe when sap drips from the pruned ends of the branches. Prune the ends of the other branches on the same trees after the leaves begin expanding and see if sap flows then. A night of above-freezing temperatures followed by a warm day does not drive sap flow in maple trees. Root pressure of maple trees is apparently not high enough to push sap from their cut branches, but can root pressure alone keep sap of some other plants flowing?

What happens after birch trees or the vines of wild or cultivated grapes are pruned at this season? High osmotic pressure in root cells drives the flow of sap in grapevines and birch branches in early spring, just as root pressure also drives the process of guttation on summer nights. The flow of sap from the cut surface of a grapevine or a birch branch can be so steady and copious that a person can collect a glass of sap within two to three minutes. This pure sap filtered through root cells is slightly sweetened with sucrose and flavored with (1) mineral nutrients such as potassium, calcium, magnesium, sodium, and manganese, (2) simple organic acids such as malic acid, citric acid, and succinic acid, and (3) a variety of amino acids. Wild grapevines can be weeds in many woods and abandoned fields; tapping their spring sap provides a refreshing, novel, and nutritious drink.

4

THE JOURNEY FROM FLOWER TO FRUIT AND SEED

At one season a plant is covered with flowers that at a later season remarkably transform into fruits. Just as a simple flower bud contains all the basic structures of a future flower (fig. 4.2), and a simple seed contains all the basic structures of an entire plant (fig. 1.2), careful inspection of each flower reveals that each contains all the rudimentary structures of the future fruit (fig. 4.7). Juicy tomatoes, crisp squash, and sweet-flavored pea pods all begin forming when male pollen grains from stamens of the flower join female pistils in an endeavor known as pollination.

Appreciating the initial origins and eventual fates of these male and female structures of flowers enriches our understanding of what actually happens during pollination

Figure 4.1 Among the flowers and stems of a squash plant, a mouse and a toad mingle with pollinators, insect predators, insect pests, and weeds. The colorful squash borer moth (*center right*) is a pollinator like the two honey bees (*center left and upper left*); however, the moth's larva bores along the length of squash stems, often leaving a wake of wilted stems and leaves in its path. The praying mantis (*upper right*) and the wolf spider (*directly below the mantis*) are stealthy—and often fast-moving—predators, like the metallic-green long-legged fly (*lower left, on the squash fruit*). A tiny midge to the right of the praying mantis has emerged from the garden soil where it (with the help of countless soil fungi; one example in the lower left corner) spent many weeks as a recycler of plant litter. The two tall weeds below and to the right of the flying midge are spiny sida—weeds in the same plant family as cotton and okra.

and fertilization of flowers—that remarkable process that makes fruits, seeds, and new generations of plants possible.

When a pollen grain, or immature male gametophyte, meets the pistil—the part of the flower containing the female gametophyte—the journey from flower to fruit begins. Figures 4.3 and 4.4 help trace

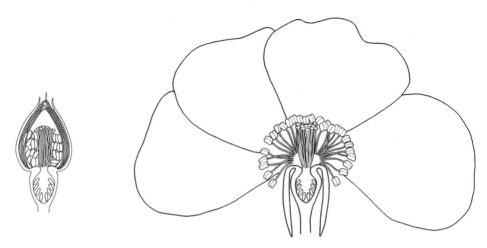

Figure 4.2 A simple flower bud (*left*) contains all the structures of the future flower. The rose flower contains male gametophytes in its stamens and female gametophytes in its pistils. The sperm cell from the male gametophyte and the egg cell from the female gametophyte join to form the first cell of the future plant.

the early steps in this long journey. As preparation for their journeys, each pollen grain in the male portion of a flower (stamen) must transform from a microspore into a male gametophyte (fig. 4.3, top); the megaspore in the female portion of a flower (pistil) must transform into a female gametophyte (fig. 4.3, bottom). Pollination requires that the grain of pollen (a microspore) first find its way to the topmost sticky part of a flower's pistil, from where it then divides to form a male gametophyte that grows until it reaches the base of the pistil, where the female gametophyte is found in a chamber called the ovule (fig. 4.4).

Each ovule of a flower is destined to become a seed, and its outer integument is destined to become a seed coat. Each female gametophyte within this ovule consists of seven cells and eight organelles called nuclei; normally every cell has a single nucleus containing all its hereditary information. However, the largest cell of the female gametophyte known as the central cell has two nuclei called polar nuclei. After pollination, these cells of the female gametophyte

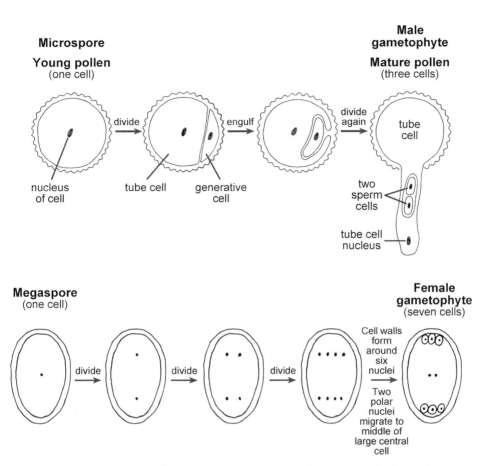

Figure 4.3 *Top:* The journey from microspore to male gametophyte occurs in the flower. The life of a pollen grain begins with one cell (the microspore) and ends with three cells, collectively known as the male gametophyte. In the drawings, nuclei of cells are represented by black dots. *Bottom:* The journey from megaspore to female gametophyte occurs in the flower. At the base of the pistil within each ovule, the megaspore begins as one cell and ends with seven cells, collectively known as the female gametophyte.

will join with cells of the male gametophyte to form the seed's embryo and the seed's nutritive endosperm, all enclosed in its seed coat.

Once attached to the sticky surface of the pistil, a grain of pollen sprouts a tube that grows all the way from the top of the pistil until

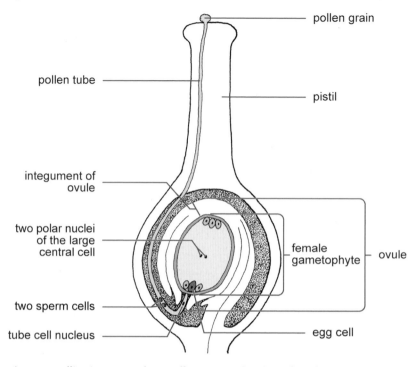

pollen grain

pollen tube

pistil

integument of
ovule

two polar nuclei
of the large
central cell

female
gametophyte

ovule

two sperm cells

tube cell nucleus

egg cell

Figure 4.4 Pollination occurs when a pollen grain reaches the surface of the pistil, and fertilization occurs when a female gametophyte meets the two sperm cells of the male gametophyte.

it fuses with an ovule at the base of the pistil (fig. 4.4). Within each young pollen grain or microspore, the single cell that initially makes it up divides asymmetrically to give rise to a smaller generative cell and a larger tube cell. The larger cell engulfs the smaller generative cell before the smaller cell divides again to form two sperm cells. The larger tube cell then pioneers the pollen's route to the ovule for the two smaller sperm cells. The tube cell's job is completed, and the jobs of the two sperm cells are about to begin.

Each pollen tube has the same three cells: the tube cell envelops the two sperm cells and transports them to the ovule. One sperm cell fuses with the egg cell to give rise to a new plant embryo; the second sperm cell joins with the two polar nuclei of the ovule's large central cell to give rise to nutritive tissue stored within every angiosperm seed, known as endosperm (*endo* = within; *sperm* = seed), and which nourishes the plant not only as an embryo but also often as a germinating seed in its first challenging days of growth. Each ovule becomes a seed only after it has fused with the two sperm cells of the pollen tube. In bean, peanut and apple seeds, the embryos use up all the endosperm well before germination of the seed (figs. 1.2 and 1.3); but in other seeds—such as those of pepper, rye, and tomato (fig. 1.4)—enough endosperm is still around at seed germination to continue providing nutrition for the new seedlings.

Pollen that finds its way to the sticky surface of a flower's pistil is either transported by wind, as in the case of corn or wheat or rye—but in most cases by flying insects, hummingbirds, or bats. However, being pollinated often involves more maneuvering than being passively transported from a stamen to a pistil. In the case of some flowers, such as those of tomatoes, eggplants, potatoes, blueberries, and cranberries, the pollen is entrapped in tubular stamens that firmly hold their pollen grains out of reach of wind and insects (fig. 4.5). Before these grains can reach pistils and begin pollinating, they first must be vigorously shaken from their stamens, and one of the best ways is through the buzzing action of bees. This special type of pollination is referred to as buzz pollination or sonication pollination. The latter term refers to the use of sound energy to clean particles and dirt that adhere to glassware, eyeglasses, and jewelry in the same way that buzzing shakes pollen loose from stamens that so tenaciously hold it. The average buzzing frequency of about 270 vibrations per second (270 hertz) vibrates the stamen so vigorously that the pollen grains are dislodged from their stamens. Not only do the bees obtain much-coveted pollen and nectar, but also the pollen of these flowers is at last free to settle down on a pistil and begin its final journey.

OBSERVE: Take a closer look at different flowers in the garden. Notice the different shapes of stamens. Many are rotund and perched

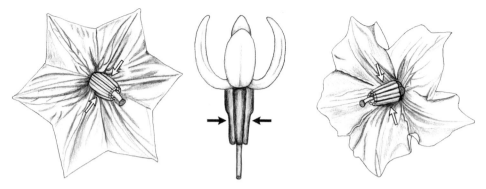

Figure 4.5 Pollen entrapped within tubular stamens (*arrows*) of some flowers must be shaken free by the buzzing of visiting bees. In each of these flowers, these stamens surround a central pistil. For flowers of some plants in the heath family (Ericaceae) such as cranberry (*center*) and plants in the nightshade family (Solanaceae) such as eggplant (*left*) and potato (*right*), buzz pollination may be the only way—or at least the most efficient way—to ensure that their pollen is dispensed and dispersed.

on long stalks called filaments (fig. 4.2 and pea flower in fig. 4.7); others are tubular and without obvious stalks (fig. 4.5 and tomato flower in fig. 4.7). Which flowers have a dusting of pollen on their stamens, and which have smooth stamens with no obvious pollen grains? Do the different butterfly, beetle, fly, wasp, and bee pollinators have a preference for which of these flowers they visit?

To reach its destination at the ovule, a pollen tube can travel as short a distance as a fraction of an inch in beet and tomato flowers or as far as a couple of feet for flowers of corn with their long silks that represent their exceptionally long pistils. The journey of the pollen tube can take only a few hours but may require several months; however, the distance traveled does not always determine the time taken to complete the journey.

The Meeting of Pollen and Pistil

OBSERVE: When squash and zucchini are at the peak of their blooming, the plants will have an abundance of flowers that are ideal subjects for an experiment that asks how pollen finds its way all the way from the top to the bottom of a pistil. A special feature of squash plants is that, unlike most garden plants, they have two types of flowers, one type that produces only stamens and pollen and the other type that produces only pistils and fruit (fig. 4.1). The first type of flower—staminate—is named after the structures that hold pollen, the stamens; and the other—pistillate—is named after the flower structures that give rise to the fruits, the pistils.

You can simulate the nurturing environment that pollen grains traverse in their journeys by preparing a special culture medium for pollen. Prepare a simple culture medium with water, sucrose, honey, a few crystals of boric acid, and a pinch of Marmite (a commercially available yeast extract). Here is the exact recipe:

2 g sucrose (cane sugar)
1.4 g honey
boric acid, 3 small crystals
smear of a concentrated yeast extract called Marmite (about 1 cm^2)
40 mL distilled water

Scatter hundreds of large yellow pollen grains from the staminate flower of a squash over the bottom of a 100-millimeter petri dish containing this culture medium. Let them settle in their new environment for about ten minutes and watch for any changes. Then, in the very center of this dish, place a flattened pistil from a female flower. Over the next few minutes and hours, observe how the pollen grains respond to the pistil in their midst.

How soon do pollen tubes begin sprouting from pollen grains? How far can they grow? Do growing pollen tubes influence other nearby tubes? Do the tubes overlap or avoid one another? How does their growth resemble the growth of the many root hairs that form on the first root of a radish seedling? How do male pollen tubes respond to the arrival of more than one female pistil? After you observe what transpires between pollen and pistil in a laboratory petri dish, follow what happens to flowers that have been pollinated in the garden as they transform into fruits.

The gallery of pollen portraits in figure 4.6 reveals a hidden beauty of plants. As John Muir once observed in a letter, "The most microscopic portions of plants are beautiful in themselves, and these are beautiful combined into individuals & undoubtedly are all woven with equal care into one harmonious beautiful whole."

HYPOTHESIZE: Pollination is usually the prelude to seed and fruit formation (fig. 4.7)—but not always. Fruits can form from flowers that have never been pollinated or fertilized and have not formed seeds. Seedless watermelons and seedless oranges sold in groceries are examples of fruits that have formed from unfertilized flowers. What allows plants to bypass pollination and fertilization on their journeys from flowers to fruits? The route taken by these seedless fruits is called parthenocarpy (*parthenos* = without fertilization, virgin; *carpy* = fruit)

Once fertilization has occurred, the transformation from flower to fruit is governed, as are almost all major events in the life of a plant, by its hormones. The familiar hormones auxin, gibberellic acid, and cytokinins are key players in this transformation. Can hormones alone set in motion the transformation from flower to fruit? If hormones alone have this transforming ability, applying an external source of one or more of these hormones at the time of flowering should cause the flower to behave as though it has been pollinated and fertilized even if it has not.

Thanks to the ready availability of commercially available auxin that is sold as a rooting powder at garden supply stores, this hypothesis is easy to test. Spray different flower clusters of tomato plants with two different solutions of rooting powder (use 100 mg/liter and 500 mg/liter). Add a teaspoon of dishwashing soap per liter along with one drop of vegetable oil (the addition of soap and oil helps the spray adhere to the flower surfaces). Spray another tomato flower cluster with water, soap, and oil solution alone. What would you

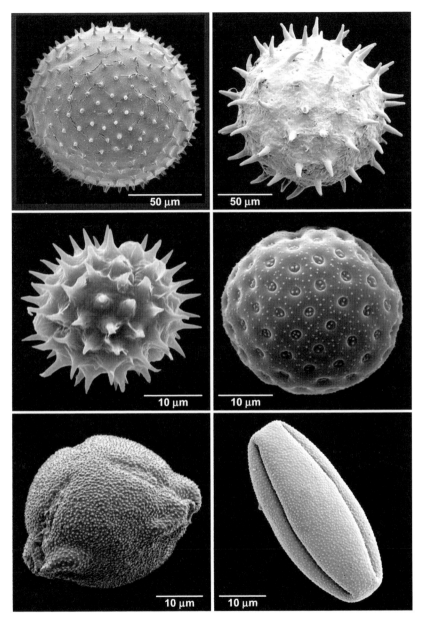

Figure 4.6 The scanning electron microscope provides images of pollen grains from six garden vegetables representing six different plant families. *Left to right, top row:* zucchini squash and okra; *middle row:* sunflower and spinach; *bottom row:* bush bean and red pepper.

The Journey from Flower to Fruit

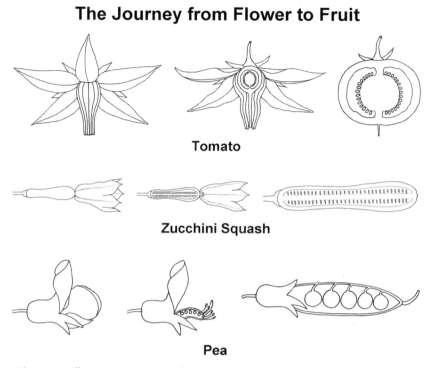

Tomato

Zucchini Squash

Pea

Figure 4.7 Pollination initiates the journey from flower to fruit. The remarkable transformation of a flower to a fruit can be appreciated by carefully observing which flower parts are lost, which parts remain but do not grow, and which parts expand and grow during this journey.

predict about the seeds and the sizes of whatever fruits form from these flowers treated with and without auxin?

The Difference between Seeds and Spores

Most plants in the garden are flowering plants and conifers that sprout from seeds, not spores. However, seeds arise from the meeting of cells descended from two spores—one male microspore and one female megaspore. How can that be? The journey of a seed is very different from the journey of a spore. The difference becomes clearer when we compare life cycles of mosses and ferns that form only spores

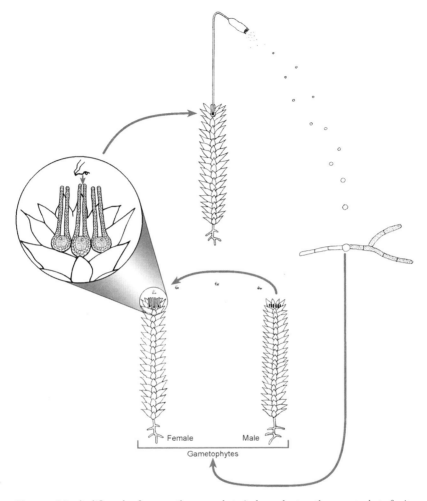

Figure 4.8 In the life cycle of a moss, the sporophyte is dependent on the gametophyte for its nutrition. The meeting of the swimming sperm and the egg shown in magnified view initiates the sporophyte generation.

(figs. 4.8, 4.9) with life cycles of flowering plants such as apples, zucchini, and tomatoes that form both spores and seeds (fig. 4.10).

Each seed holds an embryo that arises from the fusion of two gametes—an egg cell and a sperm cell. Each gamete contains only half (n) the total genetic material ($2n$) of the embryo. These gametes

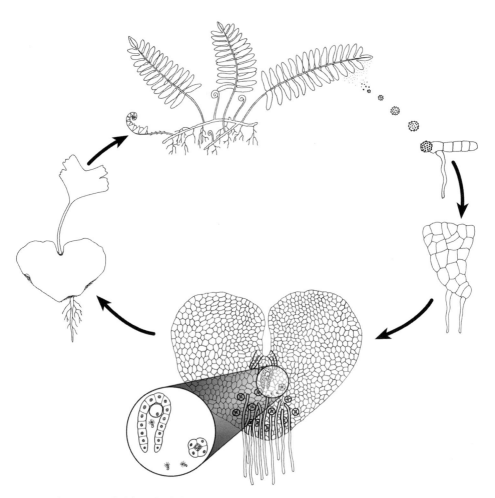

Figure 4.9 In the life cycle of a fern, the sporophyte becomes independent of the gametophyte. The fern gametophyte develops as a green plant separate from the fern sporophyte. The meeting of sperm and egg marks the birth of the sporophyte. On fern and moss gametophytes, eggs are found in structures called archegonia (*archae* = primitive; *gonia* = female reproductive organs). Corkscrew-shaped sperm swim from cellular structures called antheridia (*antheros* = male flower; *idion* = small) to fuse with egg cells of archegonia. Mosses and ferns have swimming sperm cells with flagella (*flagellum* = whip); but the two sperm cells of flowering plant pollen tubes have completely lost flagella and cannot swim. Only the flowering plants and gymnosperms have seeds with preformed sporophytes.

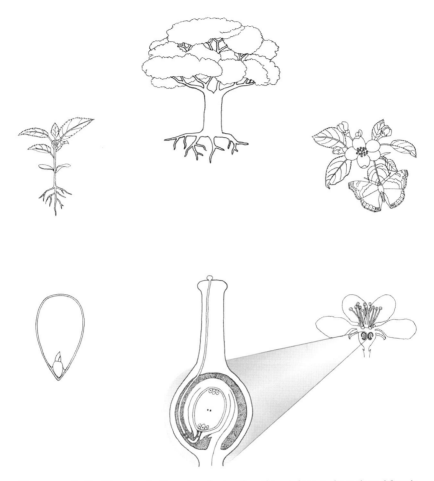

Figure 4.10 In the life cycle of a flowering plant such as this apple tree, the male and female gametophytes found inside flowers are completely dependent on their green sporophyte. The meeting of mature pollen (male gametophyte) and the female gametophyte nestled within the center of the ovule is shown in a magnified view of a portion of the apple flower. Only flowering plants and gymnosperms have flowers and seeds.

represent what is referred to as the gametophyte generation (n) of the plant's life cycle. The sperm cell is derived from a microspore and male gametophyte, and the egg cell is derived from a megaspore and female gametophyte. The embryo that results from this fusion of sperm (having genetic material n) and egg (also having genetic material n) represents the beginning of the other generation in the life of a plant—the sporophyte generation ($n + n = 2n$). The seed ($2n$) is the first member of the sporophyte generation; spores (n) are the first members of the gametophyte generation. Each seed of a flowering plant actually arises from the meeting of the descendants of two spores: one large megaspore found in the pistil and one small microspore found in the stamen. The seeds of flowering plants arise from the fusion of (a) three cells of the male gametophyte descended from the small microspore of the stamen with (b) seven cells in the center of each ovule, known as the female gametophyte, derived from the large megaspore of the pistil.

Other green plants such as ferns and mosses sprout from spores and do not form seeds. Within each seed, a future flowering plant is preformed in embryonic form; within a spore there is no hint of any organized form. The seed is made up of many cells, while a spore represents only a single cell. Few of us, however, realize that all plants that produce seeds also produce hidden spores. Every flowering plant that arises from a seed also produces microspores in the stamens of its flowers and megaspores in the pistils of its flowers. However, not all plants that produce spores produce seeds. Mosses and ferns produce neither flowers nor seeds (figs. 4.8, 4.9).

What is truly amazing is how the well-organized embryonic plant found within each seed arises from the fusion of cells that contain no obvious organization—with no hint of what will arise from these cells. Neither the microspore and its male gametophyte nor the megaspore and its female gametophyte of flowering plants and gymnosperms contain any structures that foretell their remarkable future (fig. 4.10).

Knowing When to Flower

Flowers—large, small, colorful, and plain—adorn the landscape at all seasons of the year; but each species of plant sends forth its flowers at a specific season, and not just any season (fig. 4.11). We take

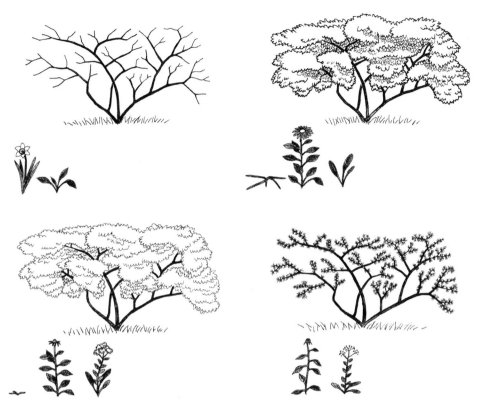

Figure 4.11 Four seasons of flowering: daffodil in spring (*top left*); purple coneflower in summer (*top right*); goldenrod in autumn (*bottom left*); witch hazel in winter (*bottom right*).

for granted that tulips flower in the spring, tomatoes bloom in the middle of summer, and chrysanthemums wait until autumn to open their flower buds. Does a plant bloom when it reaches a certain size, or a certain age? The decision of a plant to flower seems instead to be based on information that it receives about each season from its environment—the temperatures, hours of light, hours of dark, and inches of rain to which each plant is exposed. Do one or all of these features provide the information that helps make the decision for the plant? Are certain parts of plants essential for making this decision to flower?

OBSERVE: Some plants flower in the spring (such as tulips and daffodils); some plants in the summer (such as black-eyed Susan and squash); a few in the autumn (such as goldenrod and asters), and even fewer in the winter (such as witch hazel and poinsettia).

HYPOTHESIZE: To begin asking if any, or all, of this seasonal information is crucial in deciding to flower or not to flower, let's hypothesize that a plant decides to flower in response to the length of daylight, or photoperiod. All plants will be exposed to the same growing temperatures and to the same amounts of rainfall; only their exposure to light will differ. Place plants such as black-eyed Susan, goldenrod, chrysanthemum, *Coleus*, zinnia, *Salvia*, or a poinsettia from last year's holiday season under fluorescent lights that are on for sixteen hours each day or for only eight hours each day; these two exposures to light imitate respectively the exposure of a plant growing outdoors during a summer month and a plant growing outdoors during a winter month. To imitate the hours of daylight that plants would see outdoors during spring or autumn, place one of each plant species under lights that remain on for twelve hours each day. Which plants flower under which light conditions? Does the flowering of some plants seem to be indifferent to their light exposure?

Do these different plants measure a day's length by responding to the length of light, the length of dark, or both the length of light and length of dark? If plants are able to measure the length of daylight, then they should flower when exposed to that length of daylight regardless of the length of darkness. Create a light/dark cycle in which sixteen hours of light are followed by sixteen hours of dark and compare that with the normal light/dark cycle of sixteen hours of light followed by eight hours of dark. Try another light/dark cycle in which eight hours of light are followed by eight hours of darkness; compare that treatment with a normal light/dark cycle of eight hours of daylight followed by a sixteen-hour night (fig. 4.12).

To challenge these flowering plants even more, try interrupting the light periods with an hour of darkness and the dark periods with

Figure 4.12 To flower or not to flower: Do plants measure the length of day by responding to its duration of light or its duration of dark? The diagram shows how different exposures to light/dark can be used to influence a plant's decision to flower.

☀ 8 hrs.	16 hrs.		
☀ 8 hrs.	7.5 hrs.	☀	7.5 hrs.
☀ 16 hrs.		8 hrs.	
☀7.5 hrs.	☀ 7.5 hrs.	8 hrs.	

Figure 4.13 To flower or not to flower: Do plants lose their ability to measure a day's length (1) if their light exposure is interrupted by a brief exposure to dark and (2) if their dark exposure is interrupted by a brief exposure to light? Is a light interruption or a dark interruption more influential on a plant's decision to flower, or are both interruptions equally influential or equally inconsequential? The diagram shows the interruptions of light periods and dark periods that can be used to influence a plant's decision to flower.

an hour of light (fig. 4.13). Interrupt a long day of sixteen hours light/eight hours dark with a one-hour exposure to dark in the middle of the light period; interrupt a long night of eight hours light/sixteen hours dark with a one-hour exposure to light during the middle of the dark period.

HYPOTHESIZE: What happens if you graft two plants without flowers (e.g., aster or goldenrod with black-eyed Susan or purple coneflower) that are members of the same plant family (Asteraceae)? These two plants in each group happen to flower at different times of year—asters and goldenrods bloom in early autumn; black-eyed Susan and purple coneflower bloom in early summer. Use the grafting technique discussed in chapter 2 to create two mosaic plants with stems, leaves and roots of both aster or goldenrod and black-eyed Susan or purple coneflower. Earlier experiments by plant physiologists have shown that leaves of plants receive the stimulus to flower and then transmit this stimulus to the stem cells of the bud that will give rise to a flower. A plant whose leaves have been removed fails to flower regardless of the day length to which it is exposed; a plant with only one leaf can still obtain enough information from its environment to flower at exactly the right season. Some factor is induced within plant leaves by the light conditions to which the plant is exposed. This factor presumably then passes from the leaf and induces the formation of flower buds. Expose one whole mosaic plant to short days (eight hours of light) to see what flowers appear. At the same time, expose the other mosaic plant to long days (sixteen hours of light). What does the result suggest about how flowers are induced to form on different plants? Can whatever induces flowering during short days also stimulate flowering of plants that normally bloom only during long days? Can whatever induces flowering during long days also stimulate flowering of plants that normally bloom only during short days?

Try mowing plants such as purple coneflower or black-eyed Susan a few days prior to their normal blooming dates in early summer. Plants such as goldenrod and chrysanthemum that bloom in late summer can also be mowed a few days before their normal flowering dates in late summer. Would you predict that any of these plants whose intentions to flower have been stymied by the mowing will try to bloom at a later date? Once these plants have been exposed to a light/dark cycle that induces them to flower at a specific time, do they remain committed to form new flower buds and eventually bloom even if their first flower buds have been removed?

How Fruits Know When to Ripen

One rotten, overripe apple in a barrel of apples is said to spoil the whole barrel (fig. 4.14). This ripening of a fruit represents an aging process that can be compared to the aging or senescence of leaves

Figure 4.14 One rotten apple speeds up the ripening of neighboring apples and a banana.

that occurs every autumn during the massive leaf fall from trees and herbaceous plants.

Is rottenness really contagious? And if so, must the rotten apple contact the other apples or only share the same space and air with them? Does evidence exist for some substance being emitted by rotten and very ripe apples that passes on the state of rottenness to others?

OBSERVE: Start with one very ripe apple. Place one green (unripe) apple in contact with this very ripe apple; place a second green apple a foot away from the very ripe apple. Far from the first two scenarios, also place a third green apple in contact with another green apple, and a fourth green apple in contact with a very ripe banana.

HYPOTHESIZE: What do our observations tell us about some substance produced by one very ripe fruit that acts on several unripe fruits? Does separating the ripe from the unripe by a plastic barrier have any influence on the time it takes for a fruit such as a tomato, a pear, an apple, or a banana to ripen?

Place a very ripe apple in the same plastic bag with a firm green apple so that the two apples are in contact. Now repeat this arrangement of one very ripe and one green apple in the same plastic bag so that the two apples are separated by six inches. In another plastic bag, place a firm, green apple together with a very ripe apple that has been sealed inside a smaller tightly sealed ziplock bag. In another ziplock bag, place two green apples together. You can also try these same experiments with different combinations of unripe and very ripe pears and bananas.

Chemical substances produced in one location on a plant can influence tissues and cells in distance locations. To move from one location in a plant to another location, however, the chemical must be in the form of either a gas or a liquid. Solid chemicals that a plant produces, such as sugars, move from place to place only when they are dissolved in water. In some cases, the chemicals promote an action; in other cases, the chemicals inhibit an action. If rotten or ripe fruit produce some substance that speeds up the ripening (aging) of another fruit, that chemical would be considered a hormone. What chemical form—liquid or gas—does this hormone seem to take as it carries out its action? For centuries, Chinese farmers burned incense in their storage bins to promote ripening of fruits. Farmers in other lands burned cow dung for the same purpose. We now know that fumes of incense and fumes of cow dung contain ethylene—the agent responsible for fruit ripening and the only plant hormone that is known to be a gas.

This same substance—ethylene—that speeds up the ripening of fruit (and the aging of leaves as we observed in chapter 2) can slow down the growth of buds that become stems, leaves, flowers, and fruit. Remember how another plant hormone, abscisic acid, also suppresses the growth of buds in addition to the germination of seeds.

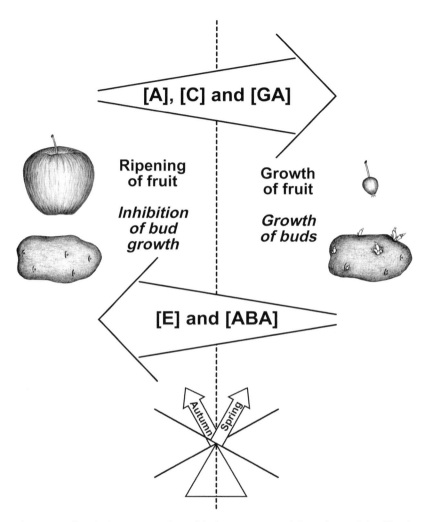

[A], [C] and [GA]

Ripening
of fruit

Growth
of fruit

Inhibition
of bud
growth

Growth
of buds

[E] and [ABA]

Autumn Spring

Figure 4.15 The relative concentrations of the hormones auxin (A), cytokinins (C), gibberellic acid (GA), ethylene (E), and abscisic acid (ABA) rise and fall within developing fruit, orchestrating its early growth and eventual ripening. The balance of hormone levels determines the fate of fruit each autumn. Likewise, the balance of hormone levels determines the fate of buds each spring. One set of plant hormones triggers the expansion of buds. However, as the levels of abscisic acid and ethylene increase each autumn, buds enter a dormant stage that will last from autumn until spring.

These two familiar hormones act in concert once again to counter the actions of other hormones.

OBSERVE: Each eye of a potato represents a bud that contains all the information to produce every part of a potato plant. Compare the influence of a ripe fruit on an unripe fruit with the influence of this ripe fruit on potato buds. If left in the dark, potatoes will soon send forth stems and leaves from their eyes; but what happens if some of these same potatoes are placed in the company of a ripe banana or ripe apple? Set up two different arrangements of fruit and potatoes and place them in dry, dark spots. (1) Place a ripe apple or a ripe banana in a paper bag with several potatoes. (2) Place the same number of potatoes in a paper bag without any apples.

As is the case for all major decisions made in the life of a plant, the same hormones—auxin, cytokinins, gibberellic acid, abscisic acid, and ethylene—appear again and again at these landmarks in a plant's life. Auxin, gibberellic acid, and cytokinins promote a plant's journey from flower to full-size but unripe fruit. Then ethylene and abscisic acid take over the task of transforming fruit from full-size but green to sweet, colorful, and ripe. It should come as no surprise that a delicate balance of these hormones governs the fate of fruit on its journey from fertilization to ripeness (fig. 4.15).

5

ENERGY FROM THE SUN AND NUTRIENTS FROM THE SOIL

The broad, green leaves of Chinese cabbage collect energy from the sun to grow even longer and wider. Plants gather energy from sunlight using their green pigment chlorophyll and convert solar energy into chemical energy. With photosynthesis (*photo* = light; *syn* = together; *thesis* = an arranging), all green plants convert the energy of sunlight into the chemical energy of sugars—a form of energy that plants and all animals can use. The process combines the energy of sunlight with the simple compounds of carbon dioxide (CO_2) and water (H_2O) to simultaneously produce sugars ($C_6H_{12}O_6$) and oxygen (O_2). Our present understanding of how plants use sunlight and chemistry to support life on Earth represents the labors of many generations of scientists.

Figure 5.1 Energy and nutrients provided by members of the cabbage family are passed directly to the green larvae of the cabbage butterfly and the many flea beetles that chew on the foliage. In turn, energy and nutrients are passed from these caterpillars and flea beetles to the mouse, the toad, a yellow jacket (*upper right*), and an iridescent ground beetle known as a caterpillar hunter (*bottom center*). Indifferent to hunting for caterpillars, a long-necked seed bug *Myodocha* (*lower right*) searches for weed seeds.

In the eighteenth century, before hardly any knowledge of photosynthesis and chemistry existed, observations of plants as they gathered sunlight helped lead the British scientist Joseph Priestley to discover not only that plants produce a so-called restorative substance when exposed to sunlight but also that this substance turned out to be a chemical element new to science. A candle burning in a closed jar soon burned out, apparently depleting the air of some substance essential for burning to occur. To prevent any surrounding air from entering the jar, Priestley inverted the large glass jar

Figure 5.2 A candle alone in an airtight container soon burns out; however, in the company of a plant in the same container, the same candle will continue burning as long as the plant is placed in sunlight, where it can carry out photosynthesis.

over a bowl of water after positioning the burning candle next to a potted plant or a living shoot from a plant (fig. 5.2). This time the candle continued to burn. He observed that "the restoration of air depended upon the vegetating state of the plant." Plants exposed to sunlight had the ability to "restore" this essential substance to air after the candle had burned out in the same upside down jar. The substance that restored air turned out to be the chemical element named oxygen. Once this new element had been discovered and its restorative attribute had been established, the discovery soon followed that the expelling of a given volume of oxygen by plants in sunlight depended on their uptake of an equal volume of carbon dioxide. By the end of the eighteenth century, the basic chemistry of photosynthesis had been experimentally confirmed, and the myriad details of this process are still being unraveled in laboratories around the world.

OBSERVE: In a bright, sunny area of a room or porch, set up a large tray of water over which you can lower a clear glass bowl or large glass container such as a small aquarium. Place a small wax candle on one edge of a platform, such as a small brick above water level; light it and lower the inverted glass container over the candle and into the water. How long will the candle burn? Now relight the candle and place it along with a small potted plant on the same platform. Does the presence of the plant "restore" the ability of the candle to burn? Is the restorative power of the plant as effective on a cloudy day?

When an aquarium containing plants of the waterweed *Elodea* is placed in bright sunlight, the leaves of *Elodea* begin bubbling rapidly. As the energy of sunlight powers photosynthesis in *Elodea*'s leaves, this submerged plant takes up carbon dioxide and water to form the sugars that the plant retains and the oxygen that the plant expels into the aquarium as hundreds of bubbles. A closed aquarium can support fish life and insect life as long as waterweeds and sunlight continually cooperate to restore oxygen to the aquarium water.

In addition to using energy ultimately derived from the sun, all living creatures need nutrients ultimately derived from the soil. The meandering roots of cabbage gather water and nutrients from far and wide, in the process concentrating nutrients from the soil eightfold. Whoever eats cabbage—such as flea beetles and the caterpillars of the cabbage butterfly—enjoys the flavor of cabbage and uses this energy and these nutrients. Whoever eats creatures that feed on cabbage—such as toads, wasps, and certain beetles—also shares some of this energy from the sun and some of those nutrient elements from the soil, now even concentrated five times more.

It takes lots of the sun's energy to grow large. The "land of the midnight sun" grows the largest cabbages on record. With sunlight powering the photosynthesis of vegetables almost twenty-four hours each day in the Alaska summer, vegetables just keep growing twenty-four hours each day. A 138.25-pound cabbage grown by Scott Robb of Palmer, Alaska, set a new world record at the Alaska State Fair in 2012. To break this record, an aspiring cabbage will have to use even more energy from the sun and nutrients from the soil. (The previous record holder, also grown in Alaska, weighed in at 125.9 pounds.)

It takes lots of energy to produce crops of fruits and nuts. Trees such as oaks, beeches, and apples produce bumper crops of nuts and fruits every two to five years. The cycle of boom and bust for trees is referred to as masting behavior, named after the old English word mast, for an exceptionally abundant fall harvest of fruit that accumulates beneath each tree. No one knows all the factors responsible for this behavior, but trees definitely expend more energy and utilize more nutrients in producing flowers and maturing fruit during mast years than they do during intervening years. Between mast years, trees can take a break from flower and fruit production, channeling their energy and nutrients into growing taller and broader.

In autumn, as a plant's exposure to light diminishes day by day, any extra energy provided by artificial lights stimulates chlorophyll production and prolongs the lives of leaves, delaying their inevitable senescence. Islands of green leaves surround streetlights in autumn; the extra light (and energy) that these leaves capture maintains their vigor as their neighboring leaves beyond the energizing influence of the lights age faster and soon fall from their trees. Meanwhile, chlorophyll in the green leaves continues to capture light energy, and that energy is channeled into sugar production. These leaves with their high content of sugars hang on the longest and often produce the brightest reds and oranges before they finally succumb to old age.

It takes energy to accomplish the work of plant growth. Plants use energy to work, to grow in height and width and depth. If the sun provides its energy to plants, then blocking some, or most, of this energy will diminish the amount of work a plant can do.

How Plants Use Light Energy to Grow

When exposed to the energy of sunlight on all sides, plants work by growing straight and green. When a plant is exposed to sun on only one side, however, it curves toward that one side. The plant elongates more on the side not exposed to sun than it does on the side facing the sun. Portions of leaves, stems, and roots that receive less energy from the sun elongate more than those parts of the plant that receive more energy of sunshine. How will an entire plant respond if it is placed in complete darkness without access to any energy of sunlight but only its own reserves of energy? Animals that do not obtain chemical energy from food provided by plants or by animals that fed on plants become weak, thin, and pale, eventually starving for lack of energy.

OBSERVE: Start three bean plants in each of three different pots. As soon as the first two leaves appear aboveground, move one of the pots to a completely dark box, cabinet, or room. Leave one pot directly under light and the third pot exposed to light on only one side. Be sure that the soil in all pots is kept moist. Check on all the bean plants after ten days and observe how growth of bean seedlings in complete darkness compares with growth of bean seedlings exposed to one-sided illumination and uniform illumination from above (fig. 5.3). Do plant parts belowground respond to the differences in exposure to light energy in the same way as plant parts that grow aboveground? Did the cotyledons shrink to the same extent on all bean seedlings during the ten-day experiment? Gently remove the potting soil from the roots of the bean plants by placing the

Figure 5.3 These three pots of bean plants were grown for the same length of time (ten days) in the same greenhouse but exposed to three different levels of light energy. The pot in the middle was placed in direct light; the pot on the right was placed in a box and exposed to light on only one side; the pot on the left was placed in a closed box with no exposure to light.

underground portion of the plants and surrounding soil in a basin with water; most of the soil should fall away from the roots. Examine the roots for any evidence that light exposure aboveground can influence plant activity belowground.

HYPOTHESIZE: What could be responsible for this elongation of certain plant parts even when they are not capturing energy from sunlight? Can you propose hypotheses that can account for why the parts of plants exposed to less light grow taller? Can you test these hypotheses? How long can plant parts continue to elongate in the dark before they eventually use all their reserves of energy?

OBSERVE: Under certain environmental conditions, some plants use more energy of sunlight in combining carbon dioxide and water to form sugars. What conditions are these?

$$\text{(photosynthesis): } 6\text{ CO}_2 + 6\text{ H}_2\text{O} + \text{light energy} \rightarrow$$
$$\text{C}_6\text{H}_{12}\text{O}_6 \text{ (glucose)} + 6\text{ O}_2$$

On hot, dry days, plants usually close their stomata not only to prevent water loss but also to restrict entry of carbon dioxide. By so doing, however, the oxygen that is trapped within the leaves acts to reverse the process of photosynthesis and diminish the production of glucose. This reversal of photosynthesis is photorespiration, a process that not only reduces the production of sugar but also liberates carbon dioxide.

$$\text{(photorespiration): } \text{C}_6\text{H}_{12}\text{O}_6 \text{ (glucose)} + 6\text{ O}_2 \rightarrow$$
$$6\text{ CO}_2 + 6\text{ H}_2\text{O} + \text{energy}$$

Some plants have overcome this inefficient loss of glucose and carbon dioxide by using additional energy and contributing additional chemical reactions that ensure all carbon dioxide is used for photosynthesis and not squandered. These plants channel all carbon dioxide into the initial step of photosynthesis. During this initial step, each molecule of carbon dioxide is first combined with a five-carbon (C5) molecule (ribulose bisphosphate) to manufacture two three-carbon organic compounds (C3 = phosphoglycerate) that are precursors of the six-carbon compound (C6 = glucose), the end product of photosynthesis.

This first step of photosynthesis is referred to as carbon fixation.

$$6\ CO_2 + 6\ C_5H_{12}O_6\text{-}2PO_3^{-3}\ \text{(ribulose bisphosphate)}$$
$$\rightarrow 12\ C_3H_6O_4\text{-}PO_3^{-3}\ \text{(phosphoglycerate)}$$

All plants carry out this first reaction of photosynthesis pathway by generating three-carbon compounds (phosphoglycerates); however, about 10 percent of Earth's plants (referred to as C4 plants) not only carry out this first step of generating three-carbon compounds but also contribute other reactions to photosynthesis that use additional energy from the sun to manufacture four-carbon organic compounds (oxaloacetate, malate) and additional carbon dioxide. Plants that carry out only the first step of photosynthesis and generate only the three-carbon compounds are referred to as C3 plants. C4 and C3 plants form the same C6 end product of glucose; however the C4 plants make more efficient use of carbon dioxide, the essential precursor of photosynthesis. In these C4 plants, carbon dioxide is first combined with a three-carbon phosphoenolpyruvate, generated by combining three-carbon pyruvate with energy and phosphate from the molecule ATP (fig. 5.4). The details of the C4 chemical reactions are presented in fact box 5.1.

C4 plants are able to carry out these chemical reactions (1C4 through 4C4) in special cells that surround the veins in their leaves without interference from oxygen and photorespiration. This organization of leaf cells is a hallmark of C4 cells that results in an accumulation of carbon dioxide in their cells that C3 leaf cells are simply unable to achieve (fig. 5.5). Within the special bundle-sheath cells found in C4 leaves, the malate produced in reaction 3C4 is cleaved into carbon dioxide, which is fixed by the C3 pathway, and pyruvate to begin the C4 cycle again, all the time minimizing loss of carbon dioxide. It takes more energy from the sun to fix carbon dioxide with

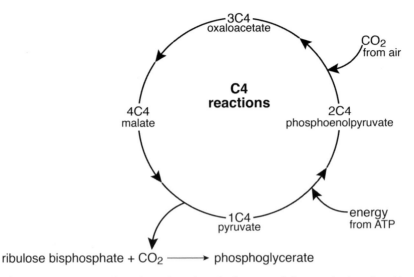

Figure 5.4 How C4 reactions channel CO_2 into the first step of photosynthesis and avoid photorespiration.

the four-carbon (C4) phosphoenolpyruvate pathway than it does by the C3 pathway. However, avoiding photorespiration and its energy-demanding reversal of photosynthesis more than compensates for this extra energy demanded by C4 plants.

OBSERVE: C4 plants outperform C3 plants under conditions of low soil moisture, high temperatures, and high light intensity. C4 plants take advantage of the extra solar energy of the intense summer sun to outcompete their neighboring plants that lack the ability to carry out C4 photosynthesis. During the hottest, driest days of summer, C3 bluegrass lawns succumb while those C4 plants such as crab-grass, spurge, and dandelion continue to thrive. C3 grasses (such as bluegrass, fescue, and quackgrass) that thrive in the cooler seasons of spring and autumn are appropriately referred to as cool-season grasses, while their relatives (such as big bluestem, crabgrass, and foxtail) that flourish best on the hot, dry days of summer are often referred to as warm-season grasses. In vegetable gardens during

---

Box 5.1 How C4 Plants Use More Solar Energy to Conserve More Carbon Dioxide

$$\text{(1C4): } C_3H_3O_3^{-1} \text{ (pyruvate)} + ADP\text{-}PO_3^{-2} + NADPH \rightarrow C_3H_4O_3\text{-}PO_3^{-3}$$
$$\text{(phosphoenolpyruvate)} + ADP + NADP$$

The transfer of energy to a molecule such as pyruvate occurs through the addition of a phosphate group from the molecules ATP or GTP (adenosine or guanosine triphosphate), known as the universal energy currency of life. To more readily follow the exchange of energy-rich phosphate groups from molecule to molecule, we can designate ATP as $ADP\text{-}PO_3^{-2}$ and GTP as $GDP\text{-}PO_3^{-2}$.

The universal electron acceptor agent in these biochemical reactions is nicotinamide adenine dinucleotide phosphate (NADP), and its electron donor agent is NADPH. Each time an electron is lost or gained in a biochemical reaction, a proton accompanies it; this pair of an electron and a proton is designated as the symbol for a hydrogen atom (H).

Three more C4 reactions (2C4 to 4C4) generate CO_2 for photosynthesis and regenerate the pyruvate that initiated these C4 reactions.

$$\text{(2C4): } CO_2 + C_3H_4O_3\text{-}PO_3^{-3} \text{ (phosphoenolpyruvate)} + GDP + NADP \rightarrow C_4H_3O_5^{-1}$$
$$\text{(oxaloacetate)} + GDP\text{-}PO_3^{-2} + NADPH$$

$$\text{(3C4): } C_4H_3O_5^{-1} \text{ (oxaloacetate)} + 2\,NADPH \rightarrow C_4H_5O_5^{-1} \text{ (malate)} + 2\,NADP$$

$$\text{(4C4): } C_4H_5O_5^{-1} \text{ (malate)} + 2\,NADP \rightarrow C_3H_3O_3^{-1} \text{ (pyruvate)} + CO_2 + 2\,NADPH$$

The carbon dioxide generated by these C4 reactions (1C4 to 4C4) is channeled into the first step of C3 photosynthesis (shown above) and used entirely to form sugars. The carbon dioxide generated by C4 plants is not lost, as it is in C3 photorespiration.

---

summer's hottest days, pigweed, purslane, and spurge grow luxuriously between the rows of beans and tomatoes. C4 weeds prosper when heat and drought stress their fellow C3 plants (table 5.1).

Where Does Soil Fertility Come From?

All life on Earth depends on energy from our sun and nutrients from the soil. Leaves and stems gather energy from the sun. From the air

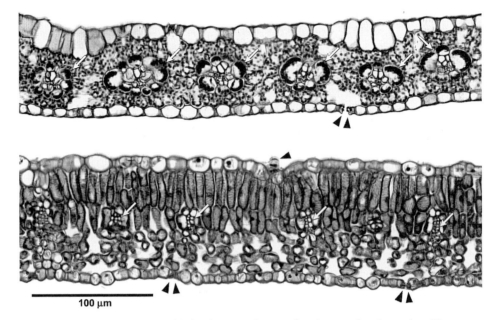

Figure 5.5 A comparison of leaf architecture for a C4 plant (corn, *top*) and a C3 plant (lilac, *bottom*) illustrates the different arrangements of bundle-sheath cells (*arrows*) surrounding the leaf veins that are responsible for the special chemical reactions in C4 leaves that maximize their conversion of CO_2 to glucose. The concentration of CO_2 remains high in bundle-sheath cells. Double arrowheads point to stomata; the single arrowhead points to a section of a trichome on the surface of the lilac leaf. See also figures 6.9 and 9.2, showing other sections of C4 and C3 leaves, respectively.

and water they gather three chemical elements—carbon, hydrogen, and oxygen—to share with roots; roots in turn gather nutrients from the soil and water to share with leaves and stems. As plants grow, almost all their mass is generated through photosynthesis using the water and carbon dioxide that they obtain from soil and air. As a plant grows in a pot of soil, neither the volume nor the dry weight of the soil diminishes perceptively. A simple experiment conducted almost four centuries ago with a pot of soil and a willow shoot demonstrated how little matter was extracted from soil as a willow tree grew from a five-pound shoot to a tree weighing 169 pounds and three ounces. By weighing dry soil in the pot at the beginning and the end of the

Table 5.1 Familiar crops and weeds can be either C4 or C3 plants

	Crops	Weeds
C4	Corn	Purslane
	Sugarcane	Pigweed
	Broccoli	Spurge
	Pineapple	Dandelion
	Cabbage	Crabgrass
C3	Potatoes	Fescue
	Wheat	Bluegrass
	Beets	Lamb's quarters
	Beans	Quackgrass
	Spinach	Cocklebur

willow's growth, Jan Baptista van Helmont (1580–1644) showed that only minuscule—almost undetectable—amounts of nutrients had been transferred from the soil to build plant tissues.

Although six of the fifteen mineral nutrients from the soil are used only in tiny amounts (macronutrients), and nine in very tiny amounts (micronutrients), each of these nutrients has been shown to be essential for the well-being of a plant. What are the sources of these nutrients—fifteen elements alphabetically arranged from boron to zinc—that come from soil, and why are they essential for the survival of plants?

Let's first consider the makeup of soils. All soils on Earth share the same three basic mineral particles of sand, silt, and clay. The relative proportions of these three mineral particles determine what is referred to as the textures of these soils. A healthy soil, however, represents a marriage of the inorganic mineral world of sand, silt, and clay with the organic world of decomposers and the decomposing matter that they return to the soil. Decomposers continually recycle those fifteen essential nutrients from the remains of plants and animals to nourish new generations of plants and animals. In the

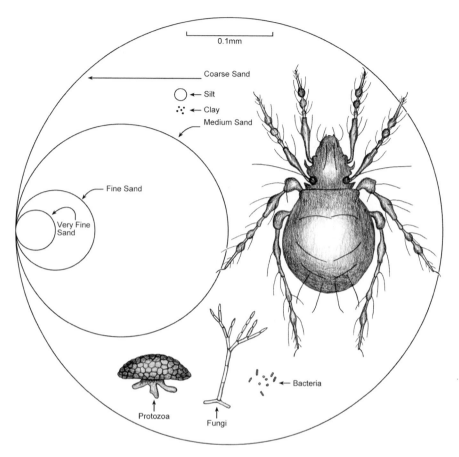

Figure 5.6 Three soil particles—sand, silt, clay—are responsible for the textures of all soils on Earth and represent the inorganic, mineral partners in a soil marriage. These particles differ in their sizes. The organic, biological partners in a soil marriage are represented by a great diversity of plants, animals, fungi, and microbes. For comparison with the soil particles of their habitat, a few of these biological partners have been included: microbial inhabitants of the soil (a fungus, a protozoan, several bacteria) and a common, small invertebrate inhabitant of the soil (an oribatid mite).

process of recycling and mixing organic matter with the three mineral particles, not only do decomposers add essential nutrients to soils, but they also add organic matter that imparts to soil a wonderful spongy structure (fig. 5.6). Decomposers physically transform an otherwise structureless soil where roots barely grow, where water drains too fast or drains too slow, into a soil through which roots, water, and air readily move. Mineral particles of sand, silt, and clay determine the *texture* of a soil, but the combination of these mineral particles and organic matter determine what is referred to as the *structure* of a soil.

The relative proportions of its large sand particles, its intermediate-size silt particles, and its tiny clay particles impart a distinctive texture to every soil. Soil textures can be coarse and sandy, sticky and high in clay particles, or silky and high in silt particles; soil textures imparted by more or less equal proportions of sand, silt, and clay particles are referred to as loams. At the end of this section, we'll see how different additions of organic matter to soils with identical textures can result in soils with very different structures—and very different fertilities.

OBSERVE: Practice preventive medicine with your plants and look for any signs that they are deficient in certain nutrients from the soil. Without these nutrients, photosynthesis languishes, cell walls are malformed, the production of essential proteins and nucleic acids suffers, conversion of light energy to chemical energy fails, and the plant shows specific symptoms of distress. What are these telltale signs that certain elements for plant growth are in short supply? In fact box 5.2 and table 5.2, these telltale signs are listed.

OBSERVE: How does application of generous amounts of synthetic fertilizer containing equal parts of nitrogen (N), phosphorus (P), and potash (potassium = K) affect the nutrition of plants? These are called balanced fertilizers; farmers and gardeners often fertilize their spring crops with these balanced fertilizers. Such fertilizers are readily available at farm and garden stores. Can you apply too much fertilizer to soil? Add ten grams of a balanced fertilizer to a small tomato plant growing in a gallon pot with standard potting soil. To another small, potted tomato plant with the same amount and type of potting soil, add ten grams of organic fertilizer in the

Box 5.2 Essential Soil Elements (Nutrients) for Plant Growth

Which fifteen soil elements are essential, why are they essential, and what are some of the signs indicating that a plant is suffering from specific elemental mineral deficiencies?

The first six elements listed here represent nutrients (macronutrients) that are required in greater amounts for a variety of different functions in plant cells. The remaining nine elements of this list are needed only in small amounts by plants; these are referred to as the micronutrients.

The alkalinity and acidity (pH) of a soil is its single most important property for influencing the uptake and availability of nutrients by plants. The macronutrients listed below are most available in soils that are neither too acid nor too alkaline—soils with pH values between 6.0 and 7.5; however, some nutrients are most readily available at high pH (> 7.5), while others are most available at low pH (< 6.0). The six positively charged micronutrient cations ($Fe^{+2 \text{ and } +3}$, Mn^{+2}, Zn^{+2}, Cu^{+2}, Ni^{+2}, Co^{+2}) are most available in acid soils. One of the few negatively charged micronutrients, BO_3^{-3}, is also most available in these soils with low pH. Unlike these other micronutrients, molybdenum, in the form of MoO_4^{-2}, is most available in alkaline soils. Not only does the alkalinity and acidity of soil affect the uptake and availability of elements, but also these elements often mutually influence each other's uptake and availability.

Macronutrients

Nitrogen (N, taken up by plants as NO_3^-, NH_4^+)

Of all essential elements, nitrogen undergoes the most movement and change, occurring in both air and soil. Nitrogen is an essential component of nucleic acids, proteins, and chlorophyll. Too much nitrate leaches calcium, potassium, and magnesium from soil.

Deficiency symptoms: Slow growth; stunting; lack of dark green color.

Good natural sources: Blood meal, fish meal, cottonseed meal, manure.

Phosphorus (P, taken up by plants as HPO_4^{-2}, $H_2PO_4^-$)

Phosphorus is an essential component of nucleic acids, proteins, and phospholipids of membranes. The energy currency of cells, ATP (a phosphorus-rich compound), is constantly synthesized during photosynthesis. Phosphorus is obviously required where growth is most intense (i.e., meristematic tissues of shoots and roots). At pH below 5.0, phosphates form insoluble compounds with Fe^{+3}; while at pH above 7.5, phosphates form insoluble precipitates with Ca^{+2}.

Deficiency symptoms: Older foliage often exceptionally dark; stunted, spindly growth; purple leaf margins of corn.

Good natural sources: Bone meal, rock phosphate.

Potassium (K, taken up by plants as K⁺)

Potassium is an essential cofactor of many enzymes and controls osmotic pressure of cells. It is an element that is essential for photosynthesis, nitrogen fixation, starch formation, and protein synthesis. Potassium helps deal with such environmental stresses as insects, drought, and winter hardiness.

Deficiency symptoms: Appear first in oldest leaves; damage and chlorosis (loss of green) at tips and edges of old leaves.

Good natural sources: Wood ashes, granite meal, greensand.

Calcium (Ca, taken up by plants as Ca⁺²)

Calcium is required during the formation of plant cell walls. Plants take up insufficient calcium if potassium levels in the soil are too high.

Deficiency symptoms: Shoot tips become flaccid; "blossom-end rot."

Good natural sources: Limestone = $CaCO_3$; eggshells.

Magnesium (Mg, taken up by plants as Mg⁺²)

Each chlorophyll molecule has one Mg atom at its center (appendix A). Magnesium not only aids in the uptake of other elements—especially phosphorus—but also serves as a cofactor of many enzymes. Plants take up insufficient magnesium if potassium levels in the soil are too high.

Deficiency symptoms: Older leaves turn yellow with green veins.

Good natural source: Dolomitic limestone = $CaMg(CO_3)_2$

Sulfur (S, taken up by plants as SO₄⁻²)

Sulfur atoms are essential components of proteins and form bonds among protein chains. Sulfur is pivotal for the production of seeds and chlorophyll.

Deficiency symptoms: Stunted plants; veins of young leaves are pale green.

Good natural sources: Gypsum = $CaSO_4 \cdot 2H_2O$; organic matter

Micronutrients

Animal manures are good, natural sources for the following micronutrients.

Manganese (Mn, taken up by plants as Mn⁺²)

Manganese is necessary for formation of chlorophyll, for processing of nitrogen, and for activating a number of enzymes.

Deficiency symptoms: Dwarfing of plants; younger leaves turn yellow with green veins; dead spots appear on leaves.

(continued)

Box 5.2   (*continued*)

Boron (B, taken up by plants as BO_3^{-3})

Boron is essential for the functioning of certain enzymes, for sugar translocation, and for promoting cell divisions. This element is also involved in the synthesis of nucleic acids and plant hormones.

Deficiency symptoms: Death of terminal buds occurs.

Iron (Fe, taken up by plants as Fe^{+2} or Fe^{+3})

Iron is necessary for formation of chlorophyll and is present in certain enzymes. Nitrogen-fixing bacteria use iron in their dinitrogenase enzymes. Iron deficiency can arise when too much soil phosphate creates insoluble iron phosphates.

Deficiency symptoms: Younger leaves turn yellow with green veins.

Zinc (Zn, taken up by plants as Zn^{+2})

Zinc is a cofactor of many enzymes; some of these are involved in production of growth hormones and chlorophyll. Excess phosphorus can limit uptake of Zn^{+2}.

Deficiency symptoms: Dead spots on leaves appear.

Copper (Cu, taken up by plants as Cu^{+2})

Copper is a cofactor of many enzymes, including those involved in chlorophyll production, formation of lignin (a tough, durable component of cell walls), and the utilization of iron.

Deficiency symptoms: Light-green leaves that become dry at their tips.

Molybdenum (Mo, taken up by plants as MoO_4^{-2})

Molybdenum is needed for the functioning of enzymes involved in nitrogen fixation and nitrogen uptake from the soil. This nutrient is most available at high pH.

Deficiency symptoms: Chlorosis; stunted growth; low yields; reddening of veins in young leaves.

Nickel (Ni, taken up by plants as Ni^{+2})

Nickel is required for the proper use of nitrogen and for the functioning of critical enzymes.

Deficiency symptoms: Pale-green leaves with cell death at the tips.

Chloride (Cl, taken up by plants as Cl⁻)

Chloride is involved in osmotic regulation, photosynthesis, and root growth. This element is absorbed in large quantities by plants and is rarely in short supply.

Deficiency symptoms: Leaf mottling and wilting; roots with stubby tips.

Cobalt (Co, taken up by plants as Co⁺²)

Cobalt is required for nitrogen fixation. Vitamin B_{12} contains cobalt.

Deficiency symptoms: Uniformly pale-green–yellow leaves. Some plants may have red leaves, petioles, and stems, or stunted growth.

form of well-seasoned (about six months old) horse manure or compost. The three nutrients of nitrogen, phosphorus, and potassium are usually well represented in manures and composts. Water these potted plants as needed throughout the summer.

At the end of the growing season, can you see any differences in the growth and vigor or the foliage and fruits of these two potted tomatoes—one nourished with balanced synthetic fertilizer and one nourished with organic fertilizer? Do either of the potted tomatoes show any of the nutrient deficiency symptoms listed above in spite of your generous addition of fertilizer—synthetic or organic—to each potted tomato?

HYPOTHESIZE: Try an experiment to track down the source of a soil's fertility. This is a classroom experiment that was first designed by the director of Britain's famous agricultural research station at Rothamsted. In *Lessons on Soil*, written in 1950, Sir John Russell presented an illuminating experiment demonstrating the source of nutrients or what he referred to as "plant food" from the soil.

Begin by digging a circular hole two feet wide and a foot deep in a portion of the garden where bare soil is exposed, trying not to mix the upper layers of soil with the bottom layers. Place soil from the upper four inches of soil (topsoil) in three pots (odd numbers 1, 3, 5) and the soil from the lowest four inches of soil (subsoil) in three

Table 5.2 A key to nutrient deficiency symptoms

Symptom	Nutrient deficiency
a. Older leaves affected	
b. Effects mostly generalized over whole plant; lower leaves dry up and die	
c. Plants light green; lower leaves yellow, drying to brown; stalks become short and slender	Nitrogen (N)
c. Plants dark green; often, red or purple colors appear, lower leaves yellow, drying to dark green; stalks become short and slender	Phosphorus (P)
b. Effects mostly localized; mottling or chlorosis; lower leaves do not dry up but become mottled or chlorotic; leaf margins cupped or tucked	
c. Leaves mottled or chlorotic, sometimes reddened; necrotic spots; stalks slender	Magnesium (Mg)
c. Mottled or chlorotic leaves; necrotic spots small and between veins or near leaf tips and margins; stalks slender	Potassium (K)
c. Necrotic spots large and general, eventually involving veins; leaves thick; stalks short	Zinc (Zn)
a. Young leaves affected	
b. Terminal buds die; distortion and necrosis of young leaves	
c. Young leaves hooked, then die back at tips and margins	Calcium (Ca)
c. Young leaves light green at bases, die back from base; leaves twisted	Boron (B)
b. Terminal buds remain alive but chlorotic or wilted, without necrotic spots	
c. Young leaves wilted, without chlorosis; stem tip weak	Copper (Cu)
c. Young leaves not wilted; chlorosis occurs	
d. Small necrotic spots; veins remain green	Manganese (Mn)
d. No necrotic spots	
e. Veins remain green	Iron (Fe)
e. Veins become chlorotic	Sulfur (S)

Note: The American Potash Institute published this simple dichotomous key for the diagnosis of eleven nutrient deficiencies in vegetables. This key enables even inexperienced gardeners to feel confident about their diagnosis of a plant ailment. Adapted from *Diagnostic Techniques for Soils and Crops*, American Potash Institute, Washington, DC (1948).

other pots labeled with even numbers 2, 4, 6. Each pot should hold one quart of soil.

Plant fifty rye seeds in one pot of topsoil (pot 1) and one pot of subsoil (pot 2). Let the rye seeds sprout and grow for four to five weeks until they attain a height of about eight inches. Then tip out the soil from the two pots and remove the rye plants—both roots and shoots. Now return the soil in which the rye plants grew to pots 1 and 2.

Each of the six pots should now be prepared for planting with mustard seeds. The topsoil of pot 1 and the subsoil of pot 2 nourished the growth of rye seedlings for five weeks before removal of the aboveground and belowground portions of the seedlings from each pot. Now the topsoil of pot 3 and the subsoil of pot 4 will each be mixed with sixty grams of fresh, shredded spinach leaves, providing raw material in these pots for recyclers to decompose. However, no plant remains will be added to the topsoil in pot 5 and the subsoil in pot 6; no raw material will be added to these pots for recyclers to decompose. The soils of pots 1 and 2 have already provided nutrients to rye plants; the soils in pots 5 and 6 have not yet nurtured any plants; and soils in pots 3 and 4 have been supplemented with plant matter.

Twenty mustard seeds should now be planted in each of the six pots. Make sure the soil in each pot remains moist but not wet. Which pot do you predict will have the largest mustard plants six weeks after the mustard seeds germinate, and which pot will have the smallest mustard plants? Is there evidence that growing rye seedlings depletes nutrients from the soil of pots 1 and 2? What do the results of the growth of mustard plants in these pots suggest about a source for "plant food" (fig. 5.7, *top*)?

To simplify the experiment, plant seeds in four pots containing soil from the same garden source (a pair of two topsoil pots—3 and 5; a pair of two subsoil pots—4 and 6), one pot of each pair having untreated soil and the other pot of the pair having soil supplemented with sixty grams of fresh, shredded spinach leaves (fig. 5.7, *bottom*).

Countless creatures of the soil—both visible creatures and the

Figure 5.7 Top: Mustard plants were grown for the same length of time and exposed to the same aboveground environment of sunlight and temperature. The plants in three pots (1, 3, 5) grew in the same topsoil; the plants in the other three pots (2, 4, 6) grew in the same subsoil. The soils in pots 1 and 2 were planted in rye for four weeks before being planted with mustard seeds. To the soil of each of the pots 3 and 4, sixty grams of fresh spinach leaves were added to the garden soil before planting mustard seeds. Mustard seeds that grew in pots 5 and 6 were planted in untreated soil from the site. *Bottom:* These mustard plants were exposed to identical environmental conditions and grew for the same length of time. As in the above experiment, topsoil was placed in the odd-numbered pots (3 and 5) and subsoil was placed in the even-numbered pots (4 and 6). The soil in pots 3 and 4 was amended with sixty grams of fresh spinach leaves before mustard seeds were planted. Mustard seeds were planted at the same time in pots 5 and 6, whose soil remained untreated.

more numerous but invisible microbes—are the ones that recycle nutrients from dead plants and animals and that make these nutrients available to living plants. The fertility that a soil provides to a plant comes from the handiwork of these living creatures—both visible and invisible—that the soil supports (fig. 5.6). According to Lady Eve Balfour, who wrote *The Living Soil* in 1943, the creatures of the soil continually convert the remains of plants and animals into humus and nutrients. By so doing, "a perfect balance between growth and decay is established, and the fertility of the soil is permanently maintained." As Sir John Russell pointed out about leaving the remains of one crop to decay, "They are not wasted, but they make food for the next crop that goes in." Nutrients are released during the decay, and humus is left. Humus is the organic matter found after all the decomposers have shredded, chewed, ingested, digested, and finally defecated whatever plant litter they have recycled. Humus particles maintain a negative charge and avidly bind mineral nutrients that have positive charges (cations). Humus breaks down slowly, remaining for years in soil and acting as a reservoir for water and nutrients, binding them and holding them within reach of plant roots. Humus is rich in the carbon and energy that remains after the energy-rich products of photosynthesis have been processed by living recyclers of the soil (fig. 5.8).

Every gardener should strive to recruit creatures of the soil as partners in the gardening adventure. These creatures help realize the ideal of fertile soil. A healthy soil represents a marriage of the inorganic mineral world of sand, silt, and clay particles with the organic world of decomposers and the decomposing matter that returns essential nutrients to the soil. The carbon-rich humus that the decomposers add to the soil imparts a spongy structure to an otherwise structureless soil. Not surprisingly, humus also holds water like a sponge and retains moisture during the hottest and driest days of summer. Without humus, soil drains too fast if it consists mostly of relatively large sand particles or drains too slowly if it consists

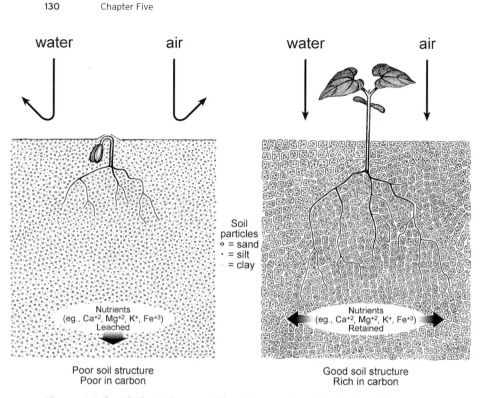

Figure 5.8 Soils with identical textures (identical proportions of sand, silt, and clay particles) can have very different structures. The carbon added to soils by the recycling of plant litter helps aggregate soil particles, allowing for the free movement of air and roots and imparting a spongy structure that retains water and mineral nutrients, shown as representative cations.

mostly of relatively tiny clay particles. The organic matter of well-decayed leaf litter that is added to soil holds many of the nutrients within reach of plants roots. A synthetic fertilizer does not contribute any organic matter to a soil. Without carbon-rich organic matter to which they can bind, the nutrients in synthetic fertilizer are quickly leached from the soil and carried beyond the reach of plant roots. Farmers and gardeners add cover crops as "green manure" to feed the decomposers. (See more about green manure in chapter 7.) The decomposers in turn not only feed the plants with nutrients that they release from the decaying cover crop, but also improve the physical environment for the plant roots as they convert green manure into humus. As decomposers mix organic matter with mineral matter, they open passageways in the soil for air, water, and other soil creatures.

Every time a crop of vegetables or weeds is harvested from a garden or farm soil, the nutrients that these plants obtained from the soil are also removed. When the remains of the crop are cleared from the land, the soil is also deprived of the benefits of this carbon-rich organic matter that acts as a sponge for water and mineral nutrients. However, when the remains of crops are returned to the land, nutrients from the soil are replenished and the essential nutrient of carbon captured as carbon dioxide from the air during photosynthesis is being returned to the soil to help restore the structure of soil and hold water and other nutrients within reach of plant roots.

We can effectively counter the global warming attributed to rising levels of carbon dioxide in the air by simply changing our agricultural practices, by promoting organic soil as an antidote to global warming. For every ton of carbon in the form of decaying plants that is returned to the soil, about three tons of carbon dioxide is removed from the air. Every organic compound in plants contains carbon. Decomposing plant litter is rich in the nutrient carbon that was taken up from the air as carbon dioxide during photosynthesis when energy from the sun converted this carbon dioxide and water to the chemical energy of sugars. Sugars then become the raw material from which other organic compounds of plants (such as the most abundant organic molecule on planet Earth—cellulose of plant cell walls) are derived. And all these organic compounds of plants fashioned originally from carbon dioxide and water become the raw material of soil carbon. The atmosphere, the soil, and all creatures aboveground and belowground benefit when the carbon atoms of the greenhouse gas carbon dioxide end up as atoms of soil carbon.

Thanks to gardening and agricultural practices of the past that have been clearly exploitative and unsustainable, our agricultural soils have been so depleted of carbon that they have lost their structure, their nutrients, and their ability to hold water. Many gardeners and farmers are now not only practicing sustainable agriculture, but they are practicing what they like to refer to as regenerative agriculture—continually adding organic carbon to the soil, not just sustaining what is already present in the soil.

How Is the Chemical Energy of Sugars Moved around a Plant?

Plants use the energy they gather from the sun to combine carbon dioxide from the air with water from the soil in the manufacture of energy-rich sugar. This sugar represents sun energy that has been converted to chemical energy. Unlike the nutrients and water that originate from roots belowground, the sugar sap of plants originates in leaves aboveground in the growing season; in winter the sugar is stored in roots. The sugary sap that drips from maple trees in the spring is under pressure generated by water moving from cells where its concentration is high to places where its concentration is lower. In the spring before leaves appear on trees such as maples, sugar has a higher concentration in roots than it does in overhead shoots; osmotic pressure moves the sap up as water is drawn into root cells from the surrounding soil. In the summer when leaves are rapidly producing sugars aboveground, sugar concentration in phloem cells is higher aboveground, so water moves by osmosis from adjacent xylem cells into phloem cells located near the cells that are actively undergoing photosynthesis and generating high concentrations of sugar. Increased osmotic pressure within those phloem cells with the higher sugar content moves the sap out to growing and developing cells that are actively unloading sugar from the phloem channels.

The stem cells of a plant's cambium continually divide to generate cells that make up the vascular system of the plant, with phloem cells forming toward the surface of the stem and xylem cells forming toward the interior of the stem (fig. 5.9). Immature phloem and immature xylem cells that are born from divisions of the cambium must progress through a series of developmental changes before

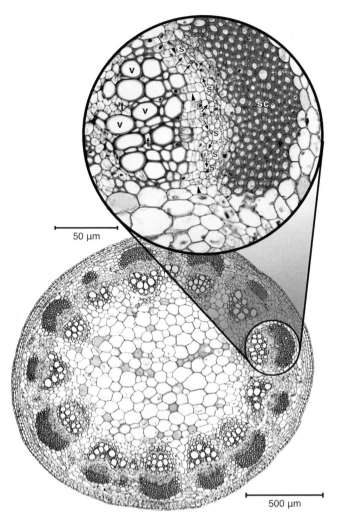

50 µm

500 µm

Figure 5.9 The thirteen vascular bundles of a sunflower stem stand out clearly in this transverse section of the stem. In each bundle the meristematic cambium cells separate phloem cells from xylem cells. *Inset:* A close-up view of a vascular bundle shows the meristematic cambium (*arrowheads*) that divides to form phloem cells to the right and xylem cells to the left. Differentiated xylem cells (vessels = v; tracheids = t) are hollow, lacking nuclei and organelles. Some differentiated phloem cells (companion cells, *arrows*) have retained their nuclei and organelles, while their sister cells (sieve-tube cells = s) have lost their nuclei and moved their remaining cellular contents to the cell's periphery to assure that sap can flow unimpeded. The rigidity of each vascular bundle is reinforced by sclerenchyma cells (sc) that have undergone programmed cell death but have retained their thick, supportive cell walls.

they attain maturity and become fully functioning members of a plant's vascular system. Phloem cells, like xylem cells, form channels of hollow cells called sieve tubes for conducting sugars and water. After dividing to form differentiated cells, xylem mother cells form hollow vessels and tracheids, both of which lose their nuclei, vacuoles, and organelles. Only the cell walls of tracheids remain and act as channels for the flow of water and nutrients. The larger vessel cells of xylem not only lose their living contents but also the cell walls separating vessel cells along a particular channel or pipeline. With the disintegration of these end walls, fluids can flow unobstructed along the length of the vessel pipeline. Unlike the vessels and tracheids of xylem tissues (fig. 3.10), however, that are programmed to die as they differentiate, phloem cells follow a different path to maturity (fig. 5.10). A phloem mother cell undergoes an asymmetric division to form a larger cell destined to become a sieve-tube cell and a smaller cell destined to become the companion cell for the sieve-tube cell. While the larger sieve-tube cell loses its nucleus and its vacuole and confines its remaining cell organelles to the periphery of the cell, its companion cell retains its nucleus, its vacuole, and all its organelles. This smaller but more complete cell apparently supports the functioning of its sister sieve-tube cell. To channel and facilitate the flow of sap, the sieve-tube cell becomes hollow and develops pores at both of its end walls. Each plant's vascular system generated by a few stem cells of the cambium provides transport of water, nutrients, and multiple compounds aboveground, belowground, and throughout the plant.

OBSERVE: Aphids can be extremely common on vegetable stems and leaves in the garden. Notice how the rear ends of these aphids often exude excess sap called honeydew. Aphids tap the phloem cells that transport sugar in the same way that we tap the trunks of maple trees to collect their sap for making maple syrup (fig. 5.11). We use hollow pegs called spiles to reach the maple sap; aphids use hollow mouthparts called stylets. If you gently remove some of these aphids

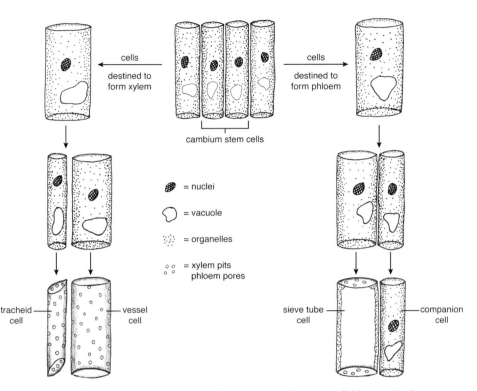

Figure 5.10 This diagram traces the birth, development, and maturity of phloem and xylem cells.

with fine tweezers, fine scissors, or a nail clipper (note: this will unfortunately kill the insect), in many cases the hollow mouthparts, or stylets, of those aphids remain behind, still embedded in the phloem cells that transport the sugar sap, acting as spiles and continuing to ooze honeydew. They resemble the spiles used to tap the buckets of maple sap that are boiled down to make maple syrup. Aphids are clean little insects, and you can try collecting a few drops of honeydew as a new tasting experience. Based on the movement of sap through aphid stylets, the rate of sap flow in phloem cells has been calculated to be as fast as 500 to 1,000 millimeters per hour. Saps differ in their sugar contents; some plant saps are under more pressure

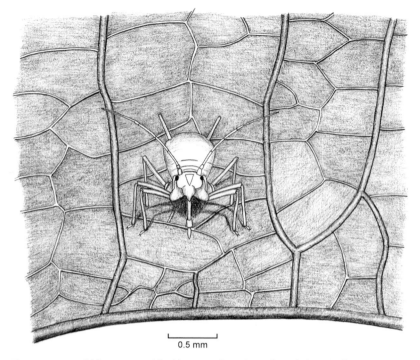

Figure 5.11 An aphid taps sugar-rich phloem sap from the surface of a bean leaf with its hollow stylets.

than saps from other plants, and aphids that feed on these sweeter saps probably dribble more honeydew.

HYPOTHESIZE: Do plant-feeding insects prefer sweeter plants — or is this not the case? Aphids derive their livelihoods from feeding on plant saps and their energy from the sugars in the saps. Should one expect aphids to prefer leaves and stems that are most active in converting light energy into the energy of sugars? Could these active, healthy plants be using some of their energy to produce substances that repel insects? How does the distribution of sweet sap relate to the distribution of aphids? How does the sugar content of sap vary within a plant?

A simple method for measuring the sugar content of sap was developed in the 1800s by a German engineer named Adolf Brix. The method uses a meter that measures the bending of light as it passes from air into water that contains different concentrations of sugars. As light passes from air into water, its path bends and the extent of bending or refraction is proportional to the concentration of sugar in the water. The Brix meter, or sugar-content meter, measures this refraction of light by a drop of plant sap. The plant sap is best obtained by using a garlic press to squeeze juice from a wad of leaves or succulent stems. Remember that as photosynthesis and sugar production progresses during the day, the sugar content of a leaf will rise, so take your measurements for comparison at the same time of day and at the same temperature.

How does the sugar content of a plant influence its appeal to plant-feeding insects such as grasshoppers, aphids, and caterpillars? Larger plant feeders such as cows, goats, and humans have insulin to handle their sugar uptake; but consuming sap with high sugar content can be an osmotic shock to the digestive tract of an insect. Remember that a high concentration of any chemical in solution will draw water by osmosis from surrounding gut tissues and blood. Water thus "pushes" its way by osmosis (*osmos* = push) into the insect's digestive tract and could result in a bellyache and indigestion as it moves from cells of the insect's gut and blood into the contents of the aphid's gut. Organically grown pastures have sweeter grasses. Farmers claim that while their livestock prefer these sweeter grasses, insects that also graze these grasses (such as grasshoppers) prefer grasses with lower sugar content. Growing organic, sugar-rich crops is a fine, nonconfrontational way to convince plant-feeding insects to move on to other pastures.

6

MOVEMENTS OF VINES AND TENDRILS, LEAVES AND FLOWERS

Plant movements were among the many natural events that captured the interest of the biologist Charles Darwin. In 1875, he wrote about his detailed observations in a small book titled *The Movements and Habits of Climbing Plants*: "It has often been vaguely asserted that plants are distinguished from animals by not having the power of movement. It should rather be said that plants acquire and display this power only when it is of some advantage to them; this being of comparatively rare occurrence, as they are affixed to the ground, and food is brought to them by the air and rain."

Darwin observed that one class of plants that climb such as pole beans "may be seen to bend to one side and to travel slowly round towards all points of the compass, moving,

Figure 6.1 Among the tangle of gourd vines, climbing tendrils constantly extend and coil. A yellow cucumber beetle (*upper left*) finds the scent of cucurbitacins—those bitter compounds produced by members of the squash family of plants—to be very enticing. Below the beetle, a white-lined sphinx moth is drawn to the scent and sweet nectar of the flowers. The parasitic wasp *Pimpla* (*bottom*) and the parasitic fly *Belvosia* (*top center*) not only survey the leaf surfaces for caterpillars on which to lay their eggs but also participate in pollinating flowers. The larva of a green lacewing (*center right*) prowls among the foliage for its favorite prey—aphids and thrips.

like the hands of a watch, with the sun." These climbing plants he referred to as "twiners" because they spontaneously spirally twine around vertical supports. Darwin made a small, fine mark on the side of the pole bean stem and then followed the position of the mark on the side of the stem as its growing tip revolved in a circle. He measured the time required for the growing tip of the plant to make a complete revolution through the air, and he showed that this natural movement occurs without ever making contact with an object. The tip of the bean plant traced a complete circle in almost exactly two hours.

Darwin also observed that another class of climbing plants are "those endowed with irritable organs which, when they touch any object clasp it." These sensitive organs "spontaneously revolve with a steady motion," and many of them "owe their origin to modified leaves." When touched, they "bend quickly to the touched side, and afterwards recover themselves and are able to act again." Once a tendril secures a foothold, it "quickly curls round and firmly grasps it. In the course of some hours it contracts into a spire, dragging up the stem, and forming an excellent spring. All movements now cease. By growth the tissues soon become wonderfully strong and durable. The tendril has done its work, and has done it in an admirable manner."

Plants move at a slow, leisurely pace. While we often go about our daily affairs at a frenetic pace, plants in the garden also move about, but at an almost imperceptible yet steady, measured pace. We can use accelerated time-lapse photography to capture the slow and graceful movements of seeds sprouting, flowers opening, vines twining, and tendrils coiling. Or we can patiently observe the trails that plants follow by marking their positions along these trails throughout the hours of a day.

Twining, Gyrations, and Calisthenics

OBSERVE: Plant the seed of a pole bean in the very center of a round pot. Once a tall, thin stalk has grown above the first two leaves, keep an eye on the tip of this stalk and the tiny new leaves that are forming there. Begin marking the position of this tip on the circumference of the pot by placing a piece of tape on the edge of the pot (fig. 6.2). Check and mark the position of the tip every fifteen minutes for the next hour. Do the tips of beanstalks always move in the same direction? Clockwise or counterclockwise?

While the growing tip of the bean gyrates over the first two leaves

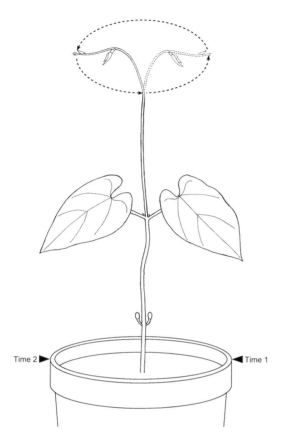

Figure 6.2 The growing tip of a beanstalk slowly spins in a circle. Positions of the growing tip at different time points are marked on the rim of the bean's pot.

Time 2 ◄

Time 1 ◄

of the pole bean, these two leaves move up and down with a regular rhythm. Who would ever suspect that an apparently placid plant moves about so much? Up and down, up and down move the leaves as they mark the passage of time with their calisthenics (fig. 6.3). Place an upright ruler or a straight stick next to, but not touching, one of the first two bean leaves. Every fifteen minutes, place a fine mark on the ruler to indicate the position of the leaf's edge. When, if ever, does the leaf reverse its direction of movement?

HYPOTHESIZE: What happens if you place a vertical pole next to the gyrating bean stalk? Do beans and other twining plants have a

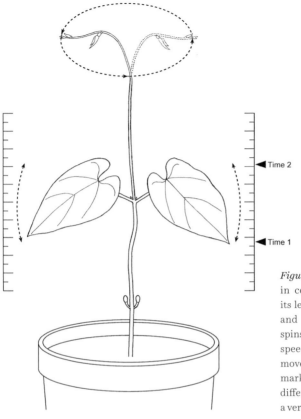

Time 2

Time 1

Figure 6.3 A bean plant is in constant slow motion: its leaves rhythmically rise and fall, its growing tip spins round and round. The speed at which the leaves move can be calculated by marking their locations at different time points along a vertical ruler.

preference for the direction (clockwise or counterclockwise) in which they spiral around a pole? What happens if you gently place a light weight (e.g., a paper clip) on the tip of one or both bean leaves that are rhythmically moving up and down? Do the bean leaves continue to move? Do they move at the same rate?

Tendrils and Touch

The other group of climbing plants moves with the aid of touch-sensitive organs or tendrils that respond to contact with an object by clasping that object. Cucumbers, gourds, and pea plants send forth tendrils that are sensitive to the lightest touch and curve toward the touched surface of the tendril.

OBSERVE: A touch as light as a loop of fine thread draped over the tip of a straight tendril (fig. 6.4) soon triggers the clasping response of the tendril, with the tendril forming coils around the thread. Plant parts bend in response to light, in response to touch, and in response to gravity.

HYPOTHESIZE: Will a straight pea tendril coil in response to a single brief touch from a finger or only to a sustained touch such as that from a loop of thread? Taking a close (microscopic) look at the bends in stems, tendrils, petals, and roots may reveal if there is a

Figure 6.4 Straight tendrils of pea plants are very sensitive to touch, even that from a loop of thread—responding to this first solid object that they contact by bending and coiling.

visible difference among the cells found on the two sides of a bend (figs. 6.5, 6.6). How would this difference among the cells account for the bending of a tendril of a pea or the stem of a pole bean? How might one or more plant hormones be involved in the coiling of tendrils?

Daily Movements of Flowers, Leaves, and Branches

Have you ever watched a dandelion or a morning glory begin and end its day? Observe how these bright flowers respond to the world

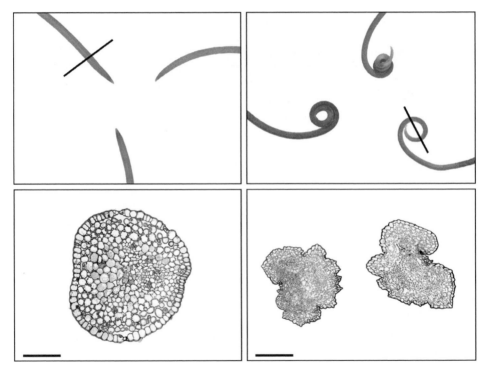

Figure 6.5 Straight tendrils of peas (*top left*) quickly transform to coiled tendrils (*top right*) in response to touch. After examining thin slices of tissue taken through these tendrils at the two locations marked by the straight black lines, we can see how the cells that make up these tendrils change during this transformation from straight (*bottom left*) to coiled (*bottom right*). Scale bars = 100 μm.

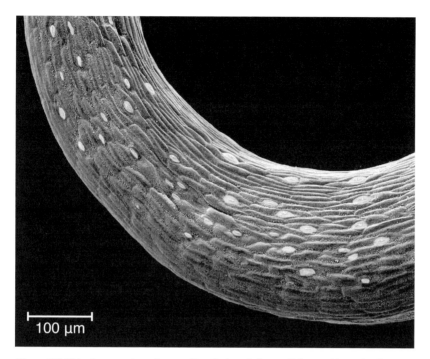

Figure 6.6 This close-up view of a gourd's coiled tendril reveals how epidermal cells on its surface all stretch with the same orientation along the length of the tendril. The presence of numerous stomata (lightly colored) among the epidermal cells belies the tendril's origin as a modified leaf. Darwin and his contemporaries pointed out that tendrils have all the features of plant organs derived from leaves.

around them. On a cold, cloudy day, dandelion flowers remain closed; only when the sun comes out will all the petals of a flower unfurl to greet the sun. Is the warmth, the light, the moisture in the air, some or all of these environmental factors responsible for how dandelion flowers behave? As each spring or summer day progresses, the open petals of morning glories close and wait for a new sunrise before opening (fig. 6.7). Four o'clock flowers (*Mirabilis*), on the other hand, depend on night-flying moths for their pollination and wait until late afternoon to open, not closing until the following morning. Flowers that open at sunrise and flowers that open at sunset respond to the same environmental cues in very different ways.

Figure 6.7 The flowers of the morning glory known as bindweed unfold each morning (*left*). As the day wears on, their trumpet-shaped flowers fold up for the day (*center and right*). The sequence of folding and unfolding is an example of flower origami.

The leaves and flower petals of many plants rise and fall in synchrony with the hours of light and dark. Because these movements are so closely coupled to light/dark cycles, they are referred to as sleep movements (fig. 6.8). During the hours of daylight, leaves assume a horizontal orientation, but as night descends, the leaves adopt a vertical position. Bean leaves and clover leaves, leaves of weeds such as wood sorrel (*Oxalis*) and velvetleaf (*Abutilon*) move up and down as daylight waxes and wanes. The leaf movements are governed, as so many plant movements are, by the movements of water into and out of the vacuoles of individual plant cells. As water moves in and the turgor pressure of a cell rises, the cell expands; as water moves out of a cell's vacuole, the turgor pressure drops and the cell shrinks. Hundreds of cells expanding in parts of the leaf translate into a leaf that lies in a horizontal plane, while the same cells contracting in these leaf regions translate into a leaf occupying a vertical plane.

Even whole trees move whole branches during their sleep movements between sunset and sunrise. At the end of each day, trees take

Figure 6.8 The same green bean plant photographed in the morning (*top*) and in the evening (*bottom*).

rests, letting their branches droop until they raise their branches the next morning. Using laser beams to scan the branches of birch trees at different times on a calm, windless day, scientists were able to precisely measure the movements of branches and showed that entire branches rise and fall throughout each day. Laser scans of whole trees consistently measured movements of about ten centimeters (four inches) each day as their branches droop at night and then rise again during the day. Changes in cell turgor pressures and the accompanying changes in cell volumes—whether exerted within the hundreds of cells of a bean leaf or the millions of cells in a tree branch—can account for the power of movements once ascribed only to animals.

OBSERVE: In a cornfield, the leaves of plants rhythmically curl and uncurl each day as the sun waxes and wanes. As the sun rises higher in the morning sky, corn leaves curl to shield their upper surfaces from the heat of the midday sun and uncurl later in the day as the sun sinks lower in the sky. This daily movement of a corn leaf helps the leaf and the plant conserve valuable water by reducing daily exposure to hot and drying sunlight. The flat corn leaf becomes a curled corn leaf when the lower surface of the leaf expands more than the top surface (fig. 6.9).

HYPOTHESIZE: What would you predict you would find if you compared a cross-section of a curled leaf with the cross-section of an uncurled leaf? In figure 6.9 (*bottom*), these two cross-sections are shown; what cellular differences do you observe that could account for leaf curling? These differences at the level of leaf cells can account for leaf curling, but how would you in turn account for the occurrence of these differences in leaf cells? Remember that plant cells with their sturdy cell walls can just as readily take up water as they can expel water; plant cells are capable of expanding and shrinking without bursting or collapsing. Shrinking and expansion of these tiny single cells translates into movements of entire tissues such as leaves, petioles, and flower petals, each of which is made up of thousands of cells.

OBSERVE: All plants transport water and the nutrients dissolved in this water from the depths of the soil to their topmost leaves. This massive long-distant movement of liquid takes place without

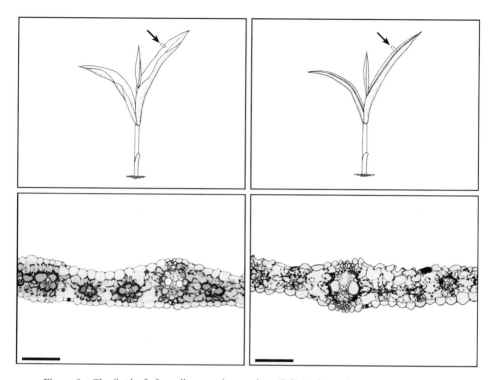

Figure 6.9 The flat leaf of a well-watered corn plant (*left*) in the early morning soon curls (*right*) on the afternoon of a hot, dry day. If we examine small, thin slices of tissue taken from these two corn leaves (*marked with small arrows and rectangles on the right leaf of each plant*), we see in the two lower panels how the cells of each leaf respond when exposed to differences in water and sunshine. Scale bars = 100 μm.

the plant expending any of its own energy. Thanks to energy from the sun, water evaporates from surfaces of leaves through many tiny pores, or stomata (*stoma* = mouth). This evaporation from leaf surfaces is referred to as plant transpiration, the botanical equivalent of animal perspiration. The water that escapes from the leaves exerts a pull on the water lower down in the plant. Water being removed at one end of a hollow straw in a glass of water pulls water from the glass into the straw and moves water upward. Each plant has multiple "straws" in the form of hollow cells arranged end to end that create channels extending from root tip to leaf tips. These hollow xylem cells form that part of the transport system, the vascular system, of plants that conduct water and nutrients upward (fig. 3.10; these cells of the plant's vascular system are illustrated and discussed at greater length in chapters 2, 3, and 5).

Those people who cut their own Christmas trees and place the cut surfaces of the trunks in a basin of water to maintain the tree's freshness discover just how much water a small tree can move from its base to its leaves—at least a liter of water each day—even after the tree has been severed from its roots. During a warm summer day, the xylem cells of a large tree channel around four hundred liters of water from soil to leaf tips.

Leafless stems of celery and roots of carrots show off their transpiration prowess when their basal ends are placed in a dye solution that moves up stems and roots, clearly highlighting their xylem channels. Although leaves and stomata are not present on these celery stems and carrot roots, their ability to transpire demonstrates that the presence of stomata is not essential for transpiration to occur. However, celery stems and carrot roots both have xylem channels that are open at their tops and at their bottoms.

Leaves vary in their number and density of stomata (fig. 6.10), but by any measure the figures are staggering. With 650,000 pores per square inch, each oak leaf has an estimated five million stomata; each corn leaf with 42,500 pores per square inch has an estimated one million stomata. These innumerable stomata act as gatekeepers through which water and oxygen are expelled from leaves, but they are also the points of entry for carbon dioxide—the essential raw material for photosynthesis. The opening and closing of the countless pores on leaf surfaces controls not only the rate at which plants transpire water from their leaves and replenish the oxygen in their environment but also the rate at which they take up carbon dioxide gas. As pointed out in chapter 5, balancing the need to prevent water loss with the need to take in carbon dioxide presents a dilemma for plants on hot, dry days, and this has been resolved in an ingenious way by certain plants that have adopted a special form

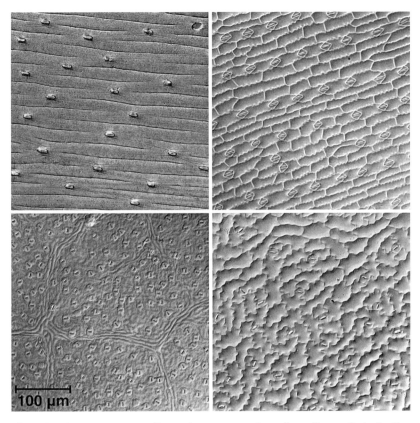

Figure 6.10 Stomata are arranged in regular patterns on the surfaces of leaves. Each plant has its own distinctive arrangement. Two guard cells flank each stomatal pore. *Monocots in top row, left to right:* Leek and corn. *Dicots in bottom row, left to right:* Oak and escarole.

of photosynthesis. The countless stomatal pores and the two guard cells that flank each pore or stoma are so important for the well-being of the entire plant that the need to close the pores to conserve water must always be balanced by the need to open pores for uptake of carbon dioxide to keep photosynthesis running.

Heat and drought are stresses that plants must face on hot, dry days. The roots that supply the water from belowground communicate any dire drought conditions to the leaves aboveground by passing a chemical signal that reduces the loss of water through the pores of the leaves. This long-distant signal sent from roots to shoots turns out to be the multitalented hormone abscisic acid, which under these environmental circumstances induces the two guard cells that open and close the stomatal pores to lose water and shrink. As the two guard cells shrink in response to abscisic acid, their pore closes, preventing the escape of additional water from the leaves and stems of the stressed plant.

Five years after publishing his studies on climbing plants, Darwin and his son Francis published *The Power of Movement in Plants*, in which they observed just how commonplace and universal these movements of plants are. "Apparently every part of every plant is continually circumnutating (*circum* = around; *nuta* = nod, sway), though often on a small scale." Plants are clearly aware of what is going on around them and have senses similar to humans and other animals. We know that they respond to touch, to light, to heat and cold, to chemicals in the air; so maybe it is not so farfetched to believe that plants even respond to the sound of music as many people claim.

7

WISDOM OF THE WEEDS: LESSONS IN HOW PLANTS FACE ADVERSITY

What makes a weed a weed? A weed just happens to be a weed because it is growing where it is not wanted; a weed encountered outside a lawn or garden is neither ugly nor harmful and is often beautiful, tasty, and useful. Weeds are members of the same plant families that we prize and nurture in our gardens. Both spinach and pigweed belong to the amaranth family; lettuce, dandelion, cocklebur, and ragweed are all members of the daisy family. But the prevailing attitude toward weeds among most farmers and gardeners is a resoundingly negative one. The ready availability and ample choice of weed killers in farm and garden supply stores reflects the generally hostile response to weeds.

If in modern times our affairs with weeds have usually evoked dislike, our ancestors adopted a more balanced,

Figure 7.1 A toad and a mouse explore the weeds bordering the garden. They share the violets, the dandelions, and the bindweed with innumerable insects, most too small to see without magnification. The bindweed flowers harbor among their pistils and stamens certain tiny insects about a millimeter in length known as thrips that thrive on the bindweed pollen. The holes in the leaves of bindweed are feeding signs of the golden tortoise beetle (*upper right*) and its larvae.

less biased point of view. In his entertaining and scholarly book on weeds, the British naturalist Richard Mabey recounts the myriad ways in which a love/hate affair with weeds has shaped the course of past human history. Our ancestors may have toiled in their agricultural fields to control weeds; however, they viewed weeds as divinely designed not only for their usefulness to humans as medicine but also for their reputed ability to foretell the future.

Although conventional opinion maintains that weeds have little, if any, redeeming attributes, weeds really serve many useful functions in a garden. Testing hypotheses about the functions of these plants of ill repute may reveal some long-held misconceptions and biases that have unjustly besmirched these unloved flowers. A book written many decades ago by the professor Joseph Cocannouer, *Weeds: Guardians of the Soil*, expounds some numerous, unsung virtues of weeds:

> protecting the soil from erosion
> improving structure of soil by loosening compacted soil with their roots
> replenishing organic matter in soil
> bringing up minerals from deep in soil with roots stretching deeper than crop roots
> nourishing and restoring soil life
> conserving and recycling nutrients that would otherwise leach away
> removing and storing carbon dioxide from the atmosphere
> encouraging biodiversity by offering habitat for creatures small and large (fig. 7.2)
> indicating soil quality

Maybe respect and coexistence offer a healthier, more economical approach for dealing with weeds. Every garden has its share of weeds, and the wise gardener welcomes these gifts and the wisdom that weeds have to offer. Living with a few weeds and understanding their ways may bring far more gardening rewards than declaring war on weeds with rototillers and herbicides.

Ex uno plura = "From One Come Many"

Most plants cannot survive being chopped, hoed, and uprooted; but for certain weeds, this harsh treatment simply improves their chance of surviving, even multiplying. Weeds with attributes of potato tubers and garlic—underground stems (rhizomes) and bulbs—each with multiple meristematic buds have the seemingly miraculous

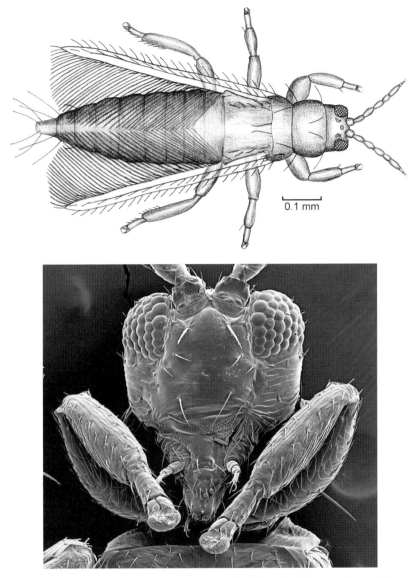

Figure 7.2 The microscope brings out the intricate but hidden facial features (*bottom*) of one of the countless thrips (*top*) that inhabit almost every flower of both weeds and nonweeds. Thrips provide nourishing meals for many insect guardians of the nearby garden, even though each of these meals appears to our unaided eyes not much longer than the period at the end of this sentence. Thrips represent one example of the rich biodiversity of insect life sheltered among the weeds. A great diversity of beneficial predators and parasites that control the activities of plant-feeding insects take up residence in the weeds.

ability to give rise to an entire new plant that can spread and thrive despite this maltreatment. Every piece of stem or bulb, large or small, that has been separated from its mother weed—as long as it has at least one bud—can sprout and begin life as a new plant. One plant can quickly become many plants: "*Ex uno plura*" (just the opposite of "*E pluribus unum*" on the Great Seal of the United States).

OBSERVE: Some weeds truly stand out for their "weediness"; three plants rank high among the weeds for their resilience and their unwavering resourcefulness. These weeds are so successful because, unlike most plants, they can regenerate an entire plant from a few stem cells that survive in fragments of stems or roots even after all plants have been uprooted, hoed under, or crushed and left to dry in the sun. Strategic placement of their stem cells enables just about any fragment of a plant to survive almost all attempts to eradicate it from a garden. After being hoed, stomped, and cultivated, plants like bindweed (*Convolvulus*), quackgrass (*Agropyron repens*), and purslane (*Portulaca*) arise undefeated and renewed like the phoenix of mythology from the chopped fragments of their parent plants.

Quackgrass goes by about two dozen common names as well as at least three scientific names—*Elymus repens*, *Elytrigia repens*, *Agropyron repens*; it is one of those plants whose place in the family of grasses Poaceae even the experts in naming plants cannot agree upon. But at least everyone agrees that the species name *repens* (= creeping) is most suitable for this grass whose underground stems or rhizomes spread by creeping underground (fig. 7.3). Personally, I prefer the name *Agropyron* for the genus of this grass, for it is this name (*agro* = field; *pyron* = fire) that well conveys the speed with which the rhizomes of this grass can "creep" across a field. About every inch or two along the entire length of its rhizome are nodes, each node having buds for roots and shoots. As long as just one node and its meristematic cells survive, this weed of many names will persist even in the face of hoeing, spraying, and pulling.

Another secret of success for the morning glory known as

Figure 7.3 The creeping rhizome of quackgrass has many meristematic nodes at regular intervals along its length.

bindweed (figs. 6.7, 7.1) lies in its underground reserves. Its dainty pink flower and delicate arrow-shaped leaves belie its robust subterranean stature. While flowers and leaves of bindweed usually stretch only an inch or two aboveground, its roots extend well over a hundred times as far belowground. Roots dive as deep as twenty feet and can spread laterally as far as ten feet in a single growing season.

Every surface of a *Portulaca* stem cut by a rake or hoe provides stem cells that give rise to roots (fig. 7.4). These roots that appear on unexpected parts of the plant are adventitious (*adventicius* = arising from outside) roots. The capability of root cells to arise from unexpected places confers remarkable resiliency to these weeds and can account for their persistence under the most adverse conditions. What meristematic region of *Portulaca* do you think provides the stem cells that divide to form these adventitious roots?

In addition to boasting the numerous but unappreciated soil-enhancing virtues of weeds listed above, weeds like violets, dandelions, pokeweed, lamb's quarters, and *Portulaca* are enjoyed for their

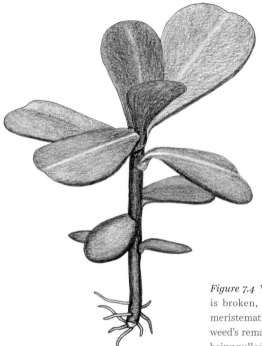

Figure 7.4 Wherever the stem of *Portulaca* is broken, adventitious roots arise from meristematic stem cells and account for this weed's remarkable ability to resurrect after being pulled, trampled, and chopped.

fine flavors and nutrient values—at least in early spring, while their shoots are tender and sweet. *Portulaca*, for example, happens to be a colorful, crisp, nutrient-dense weed that can double as a delicious and healthy garden green for salads, soups, or stir-fries not only in the spring but also throughout the summer. During his days of self-sufficient living at Walden Pond in 1854, the naturalist and writer Henry David Thoreau discovered the tastiness of *Portulaca*, known also by its common name of purslane. "I have made a satisfactory dinner, satisfactory on several accounts, simply off a dish of purslane . . . which I gathered in my cornfield, boiled and salted." Had Thoreau been aware of the recent discoveries that purslane contains high concentrations of omega-3 fatty acids in addition to its high levels of vitamins A, B, C, and E, he might have dined on purslane even more often.

HYPOTHESIZE: Weeds tell us a great deal about the condition of a soil—its fertility, its acidity or alkalinity, its wetness, its structure. The scientist Frederic Clements, known for his studies of the relationships between plants and their environments, emphasized, "Each plant is an indicator." With weeds as our guides, we can make wise decisions about improving the soil in which they are growing or about replacing weeds with vegetables that appreciate the same soil conditions.

Specific weeds provide clues about the soils in which they are growing. Many weeds—dandelions, mullein (*Verbascum* in the figwort family), docks and sheep sorrel (*Rumex* species in the buckwheat family), and plantains, for example—seem to thrive in acid soils where the nutrients potassium, phosphorus, calcium, and magnesium are in short supply in the topsoil. However, these weeds can send their roots deep enough to reach these nutrients that have leached into the subsoil and are often deficient in the topsoil. Other weeds—such as pepperweed (*Lepidium* in the cabbage family), campions (*Lychnis* in the pink family), wild carrot (*Daucus* in the carrot family)—signal that the soil is too alkaline. While spurge, common knotweed, chicory, bindweed, quackgrass, and wild mustard grow in compacted soils with poor structure, purslane, chickweed (*Stellaria*), lamb's quarters (*Chenopodium* [*cheno* = goose; *pod* = foot] in the goosefoot family), catchweed bedstraw (*Galium* in the madder family), and pigweed (*Amaranthus* in the amaranth family) thrive in very fertile, well-structured soils. Weeds can provide clues about not only a soil's supply of chemical nutrients but also the amount of organic matter and sponginess of that particular soil.

Weed control can often be accomplished by simply improving the structure of a soil, by adding healthy nutrients to soil and without adding unhealthy herbicides. Rather than using toxic, harmful, costly chemicals to control weeds, there are a variety of clever and healthy approaches to keeping weeds under control. These approaches are far more creative, rewarding, and satisfying than relying on herbicide control of a weedy situation. Since weeds are telling you about the condition of the soil in which they are growing, you can also consider digging them up and replacing them with vegetables or flowers that prefer these very same soil conditions.

Applying wood ashes not only adds a healthy dose of the essential nutrients potassium, phosphorus, calcium, and magnesium to the topsoil but also improves soil conditions for certain plants that prefer soil that is not too acidic. Applying wood ashes always reduces the acidity of soil and slightly raises its pH. Wood ashes discourage weeds that prefer acid soils. Wood ashes should be spread evenly over the soil surface before planting. About five pounds of ashes per hundred square feet provides plenty of the four mineral nutrients that are most abundant in wood ashes and that all plants need; but different plants prefer different amounts of these nutrients.

Clovers are often considered weeds in gardens. Like other members of their large plant family (which includes peas, beans, and alfalfa), clovers have nodules on their roots that are inhabited by rhizobial bacteria with the exceptional ability to take nitrogen gas from the air—a form of nitrogen that plants cannot use—and convert it to forms of nitrogen, such as ammonia and nitrates, that plants can use. (This ability of clovers and their relatives to produce their own nitrogen fertilizer is discussed further in chapter 10.) This conversion of the nitrogen gas that makes up about three-fourths of the air to ammonia and nitrates is known as nitrogen fixation. Bacteria are the only creatures that have this remarkable energy-demanding ability. Since clover roots have their own built-in sources of nitrogen nutrients thanks to their bacterial partners, they are able to thrive in low nitrogen soils that discourage the growth of other plants. Adding nitrogen-rich manure to garden soil helps these other plants compete with weedy clovers.

Like wood ashes, manure is a good source of potassium and phosphorus but also contains nitrogen and organic matter that is missing in ashes. Water represents a good percentage of fresh manure. When applying manure to the garden, add about six to seven times as many pounds (30 to 35 pounds/100 square feet) as the number of pounds of wood ashes that are recommended for the same area. Be sure that the manure is free of weed seeds by obtaining manure from stables where horses feed on alfalfa/timothy hay and pellets.

Lay out four small experimental plots in a garden. Prepare each of these at least a month before planting with vegetables. Treat the plots with the following additions and then compare the weed crops that grow throughout the spring and summer in each of them. Remove what weeds are present with hoeing before applying ashes and/or manure; but do not disturb the soil with deep tilling, such as rototilling, which brings buried seed to the soil surface and encourages their germination. Also, the many creatures of the underlying soil that collaborate in the gardening enterprise do not appreciate the violent and massive disruption of their habitat by the blades of plows or rototillers.

1. wood ashes only
2. manure only
3. wood ashes + manure
4. no additions

This series of plots tests the hypothesis that mineral-nutrient enrichment and organic-matter enrichment can discourage some or most weeds from growing in your garden.

How Weeds Deal with Their Competitors

Ecology is the study of not only interactions between organisms and their environments but also interactions among organisms, many of which turn out to be competitors with one another. Chemical ecology, in turn, is the study of the simple chemicals called secondary metabolites that mediate these interactions. Metabolites include all chemicals involved in all the chemical reactions that occur within an organism. The basic metabolites such as amino acids, hormones, and vitamins that are essential for reproduction, growth, and development of plants are referred to as primary metabolites. In addition, the elaborate chemistry kits with which plants are endowed are estimated to include around 200,000 compounds that are not essential for survival but are still important for communicating with their environments. These compounds are referred to as plant secondary metabolites and include pigments, attractants, repellents, and inhibitors, many of which—in addition to their myriad interactions with insects and other plants—are important components of our medicines and healthy diets.

Secondary metabolites mediate interactions among plants, and one secret of successful weeds is the ability of their seeds to inhibit the germination of neighboring plants with which they might compete for light, water, and nutrients. The name used to describe this inhibition of the growth of one plant by another is allelopathy (*allelo* = one another; *pathy* = harm), and the secondary metabolites involved in this inhibition are referred to as allelochemicals (appendix A).

The chemical called caffeine found in coffee and tea plants wards off human drowsiness and stimulates human activity, but has a very different role in the lives of plants. Caffeine is a simple chemical that naturally occurs in a number of plant tissues and in nature wards off attacks by insects and fungi, acting as a natural pesticide. Caffeine released by the germinating seeds of coffee and tea also turns out to inhibit the germination of neighboring plants with which they might compete for light, water, and nutrients. Any seeds that produce these inhibiting chemicals keep them safely stored away when they are not in use. In order not to inhibit their own germination, seeds that release these chemicals into the surrounding soil also produce other antidotal chemicals to inactivate whatever dose of the inhibiting chemical they might encounter.

HYPOTHESIZE: Turnip seeds—like other seeds in the cabbage family, such as broccoli and radishes—germinate within two or three days when placed on a damp surface in a warm room. They are good experimental subjects for studying the effects of naturally occurring chemicals such as caffeine on seed germination. And caffeine is a perfect chemical to test; instant coffee in grocery stores comes in both caffeinated and decaffeinated jars.

Setting up an experimental germination dish with a small amount of instant caffeinated coffee and a control dish with the same amount of instant decaf is an easy way to test the hypothesis that caffeine can inhibit seed germination. Place filter paper in the bottom of two 100-millimeter petri dishes. Dissolve a gram of caffeinated coffee in twenty milliliters of distilled water and another gram of decaffeinated coffee in a separate twenty milliliters of distilled water. Add enough of each of these solutions to moisten the filter paper in each petri dish (about five milliliters). Then sprinkle about fifty turnip seeds over the surface of each of the two petri dishes and watch for the first signs of germination.

In addition to its other weedy virtues, quackgrass is reputed to have allelopathic powers of its own. To test whether this reputation of quackgrass is deserved, compare how turnip seeds respond in the presence of different plant juices extracted with the help of a garlic press. With the press, squeeze a half teaspoon of juice from each of the following: (1) quackgrass rhizomes, (2) some dandelion stems, and (3) lettuce leaves. Add each plant juice to five milliliters of water before soaking a filter paper in the bottom of a 100-millimeter petri dish. Next, scatter fifty turnip seeds over the surface of the moist filter paper in each petri dish. Testing the influence of different plant juices on the germination of turnip seeds might be a simple and useful way to assess the allelopathic potency of any number of weeds or even garden vegetables.

Spreading Their Seeds Far and Wide

Another secret of success for weeds is their ability not only to produce many seeds but also to spread those seeds as widely as possible. One plant of pigweed, in the same family as spinach, can produce as many as 200,000 seeds in one summer. *Portulaca* might hold the record for this ability, producing up to 240,000 seeds per plant.

However, if environmental conditions for germination of weed seeds are far from ideal, all these seeds can arrest their germination and remain in suspended animation for decades. Pigweed and ragweed seeds have been awakened after a forty-year sleep; dock and evening primrose seeds have sprouted after a seventy-year slumber. Unlike the ancient *Silene* seeds of the Siberian tundra mentioned in chapter 1, most seeds cannot even come close to lying dormant for 32,000 years; but a bountiful reserve of many viable weed seeds can accumulate in the soil in only a few years.

Dispersing their seeds far and wide is a strategy for success for some accomplished weeds. Some of these weeds have come to rely on wind and water for assistance. Dandelions, thistles, and milkweeds have tufted airborne seeds that are carried aloft on currents of wind; many tiny seeds, such as those of *Portulaca* and quackgrass, are light and buoyant, carried by water currents created by rains and melting snow.

Weeds recruit help not only from the weather but also wildlife. Some plants invest energy in forming special structures for drifting on currents of air and water, for attaching to wandering creatures, and for attracting creatures who eat parts of fruits or appendages of seeds without eating the future plant or embryo of the seed. The exploding seed capsules or fruit capsules of certain weeds represent a flamboyant example of autochory (*auto* = self; *chory* = dispersal), the dispersal achieved by the plant alone. But this self-dispersal mechanism is often enhanced with the help of ants and is referred to as myrmecochory (*myrmex* = ant; *chory* = dispersal). Some seeds bear protein and lipid-rich structures called elaiosomes (*elaion* = oil; *soma* = body) that ants find quite appealing and rich in energy (fig. 7.5). Elaiosomes arise from specialized cells of seeds or the fruits in which these seeds form. As ants gather these elaiosomes as food for their brood, they leave behind the hard seeds that remain on the ant colony's refuse pile. Here in the hospitable and nutrient-rich

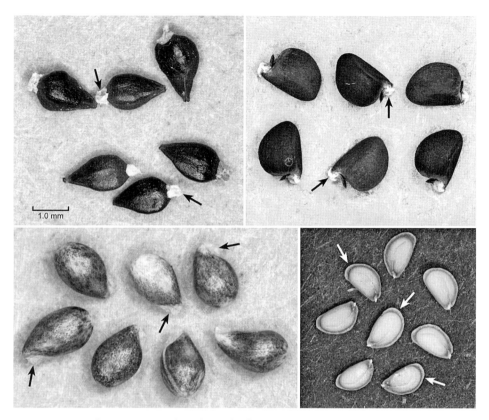

Figure 7.5 Elaiosomes (*arrows*) on the seeds of many weeds entice ants to transport these seeds to places where they probably would never have germinated if left to their own devices. *Top row, left to right:* Knotweed (*Polygonum*), spiny sida (*Sida*). *Bottom row, left to right:* Violet (*Viola*), pepperweed (*Lepidium*).

environment of the ant's home territory, the seeds find a welcoming home themselves and send down their first roots, sometimes many meters from their mother plant.

Animals of all sizes have been recruited as agents of seed dispersal. Birds, mice, and livestock eagerly eat weed seeds that then germinate in their droppings, often far from the parent weeds. Many weeds with such telltale common names as cocklebur (*Xanthium*), burdock (*Arctium*), tick trefoil (*Desmodium*), stickseed (*Hackelia*), and beggar-ticks (*Bidens*) attach to fur and clothing. Each of these seeds is equipped with minute hooks that tenaciously latch onto just about any animal surface or human clothing whose texture is neither smooth nor slick (fig. 7.6).

OBSERVE: Flinging seeds from explosive seedpods is a very effective dispersal strategy. Seedpods explode as their walls dry out and rapidly shrink in the warm sun. The motive force behind all this explosive action is the pressure exerted on seeds as water moves out of the rigid cells that surround them. As the fruit capsules begin to dry, the seeds will begin to fly. The shrinking of the walls of violet seedpods results in their seeds being suddenly squeezed and propelled from their pods. Violet seeds are fired off with great force as the walls encompassing them shrink and squeeze a seed until it is flung forth (fig. 7.7). Each of the many seeds within a seedpod of wood sorrel, or *Oxalis*, is individually wrapped in a soft, moist, elastic cellular membrane. As this membrane dries, shrinks, splits, and flips inside out, the membrane abruptly and forcefully pitches forth its seed (fig. 7.8).

Some gardens have weedy relatives of our cultivated geraniums called cranesbills. These wild geraniums are named after the shape of the central "beak" of the seedpod. This beak that resembles the long, thin bill of a crane (*geranos* = crane) actually represents the remains of the flower's pistil. Splitting, shrinking, and coiling of the cranesbill's seedpod is followed by a sudden hurling of seeds to far-off locations. First the seedpod splits into five strips; as these strips suddenly shrink and curl, they cast off the seeds that loosely cling to their sides (fig. 7.9).

The seedpods of other members of this plant family (Geraniaceae) have variations on this "bird bill" theme. Seedpods in the genera *Erodium* (*erodios* = heron) and *Pelargonium* (*pelargos* = stork) are known as heronsbills, filarees, and storksbills. Their seeds are not flung from the seedpod in catapult fashion, but instead the five seeds in each pod retain their connections to the five shrinking,

Figure 7.6 Weed seeds, by virtue of their sticky and prickly seed coats, can tenaciously grip fur or clothing of passersby, often traveling to destinations far from their mother plants. *Top row:* burdock (*Arctium*) and a close-up of burdock hooks. *Middle row, left to right:* Cocklebur (*Xanthium*), beggar's ticks (*Bidens, bi* = two; *dens* = tooth), avens (*Geum*). *Bottom row, left to right:* Several burs of stickseed (*Hackelia*) and a close-up of two burs. The burs of *Hackelia* are some of the most tenacious of burs; each projection from a bur is tipped with not just one hook, but five (*arrow*).

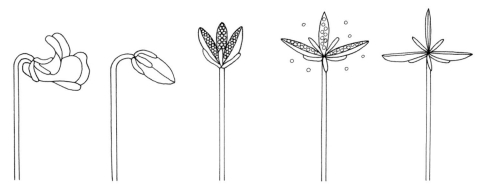

Figure 7.7 Seeds are spread far and wide by the exploding seed capsules of violets. This diagram and figure 7.8 below depict progress from flower (*left*) to exploded seed capsule (*right*).

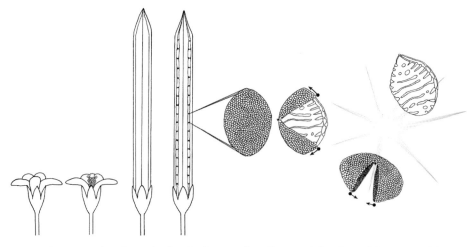

Figure 7.8 When the seeds of *Oxalis* disperse, the splitting and shrinking of the moist multicellular membrane encasing each seed (rather than the splitting and shrinking of an entire seed pod) is responsible for forcefully propelling each seed out of its pod.

splitting, and coiling strips of the pod. Each strip of the pod adopts the shape of a corkscrew. The seeds attached at the heavier end of each corkscrew-shaped strip fall to the ground; and, as changes in moisture of the air expand and contract them, the corkscrews literally screw their ways into the soil.

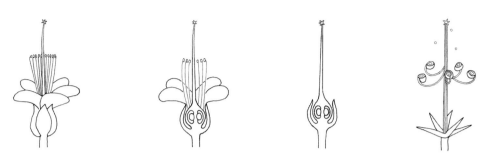

Figure 7.9 Seeds of cranesbill geraniums are spread far and wide by their exploding seed capsules. This diagram shows the progression from flower (*left*) to bursting seed capsule (*right*).

OBSERVE: Gather together several fruits (seedpods) of violets and cranesbill geraniums along with fruits of *Oxalis*; *Oxalis* is the weed commonly known as pickleweed, sour grass, or wood sorrel. Place these seedpods with their stems upright in a small beaker. Position the beaker in the middle of a large white sheet and see how far the seedpods can fling their seeds and which plant flings seeds the farthest. Make sure the stems supporting the fruits are erect and that the seedpods project well above the rim of the beaker. If you place a lightweight plastic cup over the beaker, you can listen to the occasional pings of the seeds striking the plastic. Over the next hours and days, observe how far the seeds are flung from their stems.

Some enterprising weed seeds have adopted more than one dispersal strategy. Violets are a good example of plants whose seeds not only are explosively propelled from their seed capsules but also are coveted for their delectable elaiosomes. Violet seeds provide their own dispersal and also rely on ants to carry them beyond the limits of their self-propulsion. Clearly violet seeds have doubled the assurance that they will be well dispersed — no wonder violets are so widespread and form such dense, attractive carpets of blue in lawns that are free of herbicides.

HYPOTHESIZE: Weeds should be used as allies rather than as enemies in the garden. With their deep roots, many weeds bring up nutrients from deep in the soil that many vegetable roots cannot reach. And with their spreading fibrous roots, they help loosen hard soil, blazing a trail for vegetable roots that follow in their wakes. When nutrient-rich weeds are added to the garden as mulch, they add these nutrients where roots of vegetables can reach them and help hold moisture and improve the structure of soil. Their remains provide habitat not only for decomposers that keep the soil in good shape but also for predators that keep down the numbers of plant-eating pests.

Discover for yourself how the wisdom and virtues of weeds can be used to improve a garden. Tilling the soil awakens dormant seeds; it is best to let "sleeping" weeds lie deep in the soil by avoiding tilling the soil more than is necessary. Tilling also disrupts the lives of the countless creatures that live in the first few inches of soil. These creatures of the underground are constantly improving soil quality by mixing, aerating and recycling nutrients.

Cutting biennial or perennial weeds at just the right time during their growing season—just as they have attained their maximum growth and are beginning to flower—can truly discourage their reappearance the following year. By cutting weeds to the ground after they have devoted so much energy and channeled so many nutrients into flower buds and vegetation aboveground, the remaining weed stump and its roots are deprived of the energy and nutrients that normally are returned to roots at the end of the growing season. Resources stored underground through the months of winter enable biennial and perennial weeds to resprout vigorously in the spring. By removing these weeds as they begin flowering, their seed source is eliminated and the remains of the weeds decompose, contributing nutrients and organic matter to the garden soil. The weeds are deprived of resources that they would have used for their own resprouting, but their decomposition passes on these resources to the soil.

Weeds can also be controlled far more effectively and in a far more environmentally responsible fashion by adding organic amendments to untilled soil than by tilling and spraying with herbicides. Try leaving a large section of your garden soil untilled; instead, completely cover the soil and any exposed weeds with a variety of organic amendments (*emendare* = improve). Are any or all of these methods of weed control on untilled soil more effective than attempts to control weeds in tilled garden soils? Weeds and their seeds will be smothered under a few inches of each amendment. Let the innumerable creatures of the soil slowly and gently mix this organic matter with the underlying mineral particles of sand, silt, and clay. These creatures till the garden without bringing weeds seeds to the surface. After they have been at work for several weeks or several months, depending on the season, they have prepared an inviting, spongy bed for your vegetable seeds. With only a little hoeing and raking, the soil is now ready for planting; and most of the neighborhood weed seeds remain slumbering beyond the reach of hoe and rake.

Cover crops, or "green manure," are annual crops that are sowed throughout the year—in spring, summer, and fall. These crops grow quickly, outcompeting weeds and adding plenty of organic matter to the soil. The dense soil cover that they provide also crowds out weeds that try to establish a foothold in a parcel of ground. As winter crops, they protect the soil surface from wind erosion. Their roots loosen compacted soils, bring up minerals from deep in the soil, and—in the case of nitrogen-fixing legumes such as alfalfa, peas, beans, vetches, and clovers—add a good dose of nitrogen fertilizer to the garden. Cover crops in the cabbage family such as turnips and oil-seed radish produce secondary metabolites known as glucosinolates (mentioned in chapter 9). Not only are glucosinolates and their derivatives healthy additions to our diets, but they also happen to have allelopathic attributes that probably inhibit germination of many weed seeds. While they are certainly health enhancing in human diets, glucosinolates released by cover crops turn out to be toxic to a number of invertebrate pests. Like other secondary plant metabolites, glucosinolates can double as both allelopathic agents and plant defense chemicals (as we'll see in chapter 9). Farmers, gardeners, and scientists continue to discover excellent reasons to include cover crops in their annual planting agendas, trying to minimize

Table 7.1 Common cover crops, the seasons when they can be planted, and the amount of seed to sow per 1,000 square feet

Name of plant	Season of sowing	Density of sowing (lb per 1,000 square feet)
Alfalfa*	Early spring through late summer	0.5
Barley	Early spring to summer	2
Buckwheat	Spring to summer	2–3.0
Crimson clover*	All seasons	0.7
Sweet clover*	Spring to summer	0.5
Pearl millet	Summer	0.25
Mustard	Spring to summer	0.25
Oat	Spring to summer	4
Field pea*	Spring or fall	3
Oilseed radish	Late summer	1
Winter rye	All seasons	4
Soybean*	Spring to summer	4
Sunflower	Spring	0.25
Turnip	Spring or late summer	0.25
Hairy vetch*	All seasons	1
Spring wheat	Early spring	4

Source: Adapted from the catalog of Johnny's Selected Seeds
Note: The nitrogen-fixing legumes are marked with asterisks. Those crops that are particularly efficient at suppressing weed growth are bold.

exposure of bare ground by ensuring that the soil is always occupied by some crop (table 7.1).

Seed mixes of these legumes and seeds of numerous other non-legume crops such as buckwheat, rye, mustard, oilseed radish, and sunflowers are available in garden stores and catalogs. By growing a mix of cover crops, the talents of a number of different cover crops can be simultaneously put to use in enriching the soil and controlling weeds and even certain pests. Before planting a spring or summer crop, the mature cover crop is stomped and mashed down so that creatures of the soil can begin working their magic of converting the remains of the cover crop to soil organic matter. The cover crop is now considered a green manure and will soon decay, blending its nutrients and organic matter with the mineral soil below. With a little hoeing and raking, the soil will be ready to plant.

Horse manure is a fine amendment to garden soil. I add several loads of horse manure to my garden each winter, but avoid horse manure that contains weed seeds. Horses that are fed alfalfa hay and/or timothy hay produce manure that has few, if any, weed seeds. Cover the soil with three to four inches of manure and let the soil creatures put the final touches on preparing the bed for your vegetable seeds. In a few weeks or months, depending on the season, the soil creatures have gently tilled and blended the manure with the underlying mineral soil, thus accomplishing a task worthy of a master gardener.

Autumn leaves are wonderful mediators of weed control whose talent for also enriching the soil usually goes untapped. The leaves we rake in the autumn most often go up in smoke or to landscape recycling centers. By first shredding leaves before adding them to the garden soil, you help the small decomposers of the soil in their job of not only chemically enriching the soil with nutrients but also physically enhancing the spongy structure of soil. Let the leaves decompose to fine fragments the following spring before hoeing and raking in vegetable or flower seeds.

Grass clippings are another usually wasted source of nutrients and an unappreciated agent of weed control. During the warm days of summer when lawns are mowed weekly and clippings abound, this green manure can be added between existing rows of vegetables to smother whatever weeds are growing. In addition, some grasses, such as bluegrass, are reputed to release allelochemicals that suppress the germination and growth of neighboring weeds. The grass clippings will probably be even more effective when added to the soil after those weeds have been first scraped out with a hoe and then left to dry in the sun. Like other green manures, grass clippings decompose and mix with the underlying mineral soil more rapidly than dry autumn leaves; green plant matter provides more nitrogen-rich material for the nourishment of the countless decomposing mi-crobes than any dry plant matter.

These experiments test the hypothesis that controlling weeds in our gardens is simpler and more successful when we work in part-nership with nature. The experimental results challenge the conven-tional hypothesis that successful, profitable gardening and farming require that we confront nature with synthetic pesticides, herbi-cides, fertilizers, and rototillers.

8

PLANT COLORS

The dominant green color of plants is imparted by molecules of chlorophyll—a molecule that not only transmits green light but also gathers the energy of red and blue light to power photosynthesis. This green world of plants is embellished with a rainbow of colors that can be arranged in countless combinations that are so pleasing to our eyes. There are even colors that insects and birds can see but that we cannot. The many colors that we see arise from pigment molecules that are produced only by plants; each of these plant pigments absorbs the colors we do not see and transmits the colors we actually do. These are molecules that human bodies cannot produce, but many of them are essential vitamins for our nutrition; and some protect us from the detrimental effects of various toxins in our environment.

Figure 8.1 A mouse and a toad both eye the juicy grasshopper that has hopped into this row of colorful Swiss chard. Woolly bear caterpillars (*lower right and far left*) dine on a variety of garden weeds such as dandelions and *Oxalis*. In the coming year, these caterpillars will transform into tiger moths, each visiting and pollinating many flowers during their lives as moths. A leaf-footed bug (*upper right*) uses its sharp beak to suck sap from weeds and vegetables alike. A yellow jacket (*to the right of the caterpillar at far left*) flies through the stalks of chard searching for insect meals, but the funnel-web spider (*lower right*) simply waits for an insect to stumble upon its web.

The panoply of colors and patterns with which flowers, fruits, and leaves are adorned convey beauty to a landscape and divulge information about the alluring tastes and health benefits that their plants offer. As tomatoes, peppers, and apples ripen in late summer and as leaves grow old in autumn, red, orange and yellow pigments replace the green molecules of chlorophyll. These bright and stunning colors catch our eyes and remind our taste buds of the delicious flavors we associate with the harvests of summer and autumn. Plant colors appeal to our sense of beauty, and colors of plants often tell us a lot about the nutrients and healthy substances that a plant contains.

The Variety of Plant Chemicals That Give Plants Their Many Colors

OBSERVE: Few vegetables come in as many vibrant colors as Swiss chard, whose bright colors are imparted by pigments known as beta-lains. Swiss chard, beets, four o'clocks, cacti, and *Portulaca* are about the only plants in the garden that produce these special red, orange, and yellow pigments. What these plants have in common is that they are all closely related; they are all members of related plant families. These betalain pigments act as substances known as antioxidants that protect our cells from other substances in the environment—called oxidants, or free radicals—that chemically alter and damage molecules that make up our cells. These relatives of spinach are loaded with all the nutrients of spinach and more.

Other plants of the garden such as red cabbage, red peppers, red tomatoes, and red sweet potatoes have very different red, blue, and purple pigments, called anthocyanins (*anthos* = flower; *cyanos* = blue), and carotenoids (*carota* = carrot) that produce yellow-orange colors. Although they share some of the same colors as betalain and are also potent antioxidants, they are built of different atoms arranged in very different configurations (fig. 8.2, appendix A).

OBSERVE: Vegetables may have the same colors, but different vegetable pigments with different chemical properties can produce similar or identical colors. Compare the colors of red pepper and red cabbage with those of red Swiss chard and red beets (fig. 8.3).

HYPOTHESIZE: Chemical similarities and differences between the red of cabbage and the red of beets or Swiss chard can be easily demonstrated by comparing how cabbage juice and beet juice change color when treated with various acidic or alkaline solutions

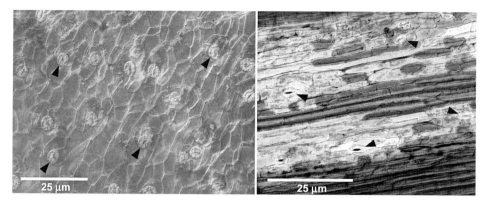

Figure 8.2 Pigment cells of red cabbage (*left*) and Swiss chard (*right*) both contain pigments that are red but that have very different chemical structures. Some stomata of these leaves are marked with arrowheads.

such as white cider vinegar or a solution of ammonia. The measure of a solution's acidity and/or alkalinity is known as its pH. Prepare juice from red cabbage and red beets by placing them separately in a blender and straining the juice. Place the two juices in large, sealed containers that can be stored in a refrigerator for several weeks; and during this time, each can be dispensed into a series of small tubes to each of which an equal volume of solution of known pH can be added. Plot the color changes that occur in cabbage juice and beet juice as they are exposed to solutions of different pHs. Which pigment can change into the most colors?

Leaves and Fruits That Change Color in Response to Light

The tiny packets, or organelles (*organ* = organ; *elle* = little), of green pigment within leaf cells are called chloroplasts (*chloro* = green; *plast* = form). Although plant cells with their stiff cellulose walls are capable of little if any movement, their chloroplasts are free to wander about inside a plant cell and shift their positions within the cell to protect them from overexposure to intense light. Under high-intensity light, chloroplasts therefore move from positions throughout leaf cells to the sides of leaf cells that are parallel to the incoming beam in order to minimize their exposure to the intense incoming light. Under low-light conditions, in shaded environments, the chloroplasts then settle back to positions throughout the cells and sometimes even orient perpendicular to the incoming light beam in order to increase their energy uptake.

Chloroplasts not only move about in cells but also undergo changes in time. The two classes of pigments that reside in chloroplasts—chlorophylls and carotenoids—are not soluble in water and lie within the hydrophobic lipid membranes of chloroplasts. Chloroplasts of fruits can undergo transformations as they ripen; red and yellow carotenoids replace green chlorophyll in chloroplasts of the pigment cells of fruits. Green chloroplasts of unripe fruits become orange, red, or yellow chromoplasts of ripe fruits (fig. 8.4).

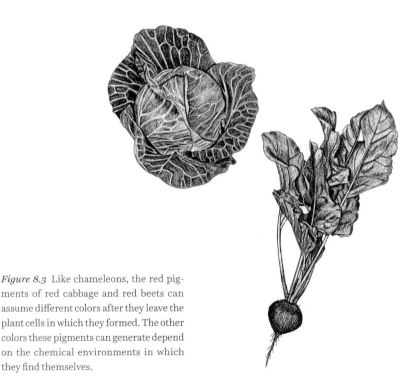

Figure 8.3 Like chameleons, the red pigments of red cabbage and red beets can assume different colors after they leave the plant cells in which they formed. The other colors these pigments can generate depend on the chemical environments in which they find themselves.

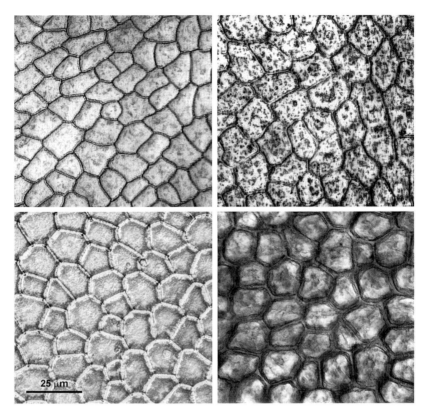

Figure 8.4 In the pigment cells of green pepper and red pepper (*top left, right*) and green tomato and red tomato (*bottom left, right*), the initially green chloroplasts transform to red chloroplasts, referred to as chromoplasts. Green chlorophyll pigment is concentrated in the membranes of their chloroplasts. As tomatoes and peppers ripen, red carotenoid pigment is likewise localized within membranes of chloroplasts that have transformed to chromoplasts. In the pigment cells of tomato, the well-known red lycopene pigment is a carotenoid.

OBSERVE: Take a leaf from a plant and place it on a wet paper towel in the top lid of a petri dish. *Coleus, Philodendron,* mustard, and spinach leaves with uniform green pigmentation are good leaves to use for this demonstration of chloroplast movements. On top of the leaf, place a high-contrast black-and-white negative, a strip of aluminum foil, or a transparency with an opaque black pattern. Now place the bottom half of the dish on top of the leaf. The leaf, paper

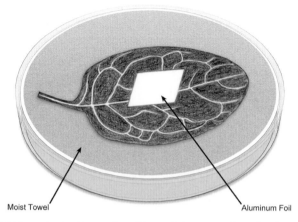

Moist Towel   Aluminum Foil

Figure 8.5 A high-contrast pattern such as this parallelogram cut from aluminum foil is sandwiched between the lid of a petri dish and a green spinach leaf sitting on moist towels before exposure to bright light. This arrangement demonstrates how light intensity can influence the rearrangement of green chloroplasts within plant cells.

towel, and high-contrast pattern are now sandwiched between the top and bottom of the dish. Press the two lids together and place the dish either in a horizontal position in bright sunlight for forty-five minutes or in a vertical position two feet in front of the bright light from a slide projector. When the leaf is removed from its sandwich, you should see an image on the leaf created by movements of chloroplasts. After removing the transparency or opaque pattern that obstructed light from reaching chloroplasts of the leaf, leave the leaf in its moist chamber either in the light or the dark and see how long the image persists (fig. 8.5).

OBSERVE: Some leaves are really ideal for observing the movements of chloroplasts within single plant cells. The leaves of garden vegetables have relatively thick leaves in which cells are arranged in several layers, making clear viewing of their individual cells difficult, if not impossible. A common waterweed called *Elodea*, however, has very thin leaves only two cell layers thick, and whose individual cells can be easily and clearly viewed with a small microscope (fig. 8.6). These can be collected outdoors or obtained from a store offering aquarium supplies. Mount one or two of its many leaves in clear pond water on a glass slide, and sandwich the leaf cells between the slide and a thin cover glass. Within the individual cells, you should be able to see individual packets of bright green. Each of these round green packets is a chloroplast. Under the bright light of the microscope, what do you think will happen to these chloroplasts? Wait and watch carefully as soon as the microscope light goes on and as soon as the chloroplasts are in focus to see if there is any action in these green leaf cells.

HYPOTHESIZE: How do you suppose these chloroplasts move about in a cell? Do they move on their own with help from another structure or some other structure(s)—visible or invisible—within the leaf cells of this pond plant? Based on what is known about the microscopic structure of plant cells (fig. I.5), can you hypothesize what might be moving these organelles from place to place in a cell?

OBSERVE: Every autumn the forested landscapes of Earth undergo stunning color transformations. The green pigment chlorophyll that has captured the energy of photosynthesis throughout the summer in countless green leaves is replaced with a kaleidoscopic combination of red, orange, and yellow pigments. Chlorophyll molecules— and the orange and yellow pigments called carotenoids—are not soluble in water and are found in membranes of the cell's chloroplasts. Autumn colors attributed to carotenoids are simply unmasked by the loss of green chlorophyll from the chloroplast membranes that these two types of pigments co-inhabit.

Violet, red, and orange colors conferred by water-soluble anthocyanins, however, are also produced in the stunningly colorful leaves of autumn trees. While the chlorophyll and carotenoid pigments are found in the membranes of chloroplasts, anthocyanins—like their equally colorful counterparts the betalains—occupy the water-filled vacuoles of plant cells. Each anthocyanin pigment is usually coupled

Figure 8.6 Most plant movements are relatively slow and occur within minutes and hours, showing up best in time-lapse images. However, the green chloroplasts in the living leaves of the waterweed *Elodea* shuffle so rapidly from place to place in their cells that their movements are measured in seconds and minutes. Arrowheads have been added as landmarks to help compare the movements of chloroplasts between time = 0 minutes (*top*) and time = 3 minutes (*bottom*).

to one or more sugar (glucose) molecules that increase its solubility in the water-filled vacuoles of cells. Any increase in anthocyanin-sugar complexes within a cell's vacuole naturally increases its turgor pressure as water readily moves into the cell by osmosis. In addition to contributing to the colors of autumn leaves, anthocyanins appear in early spring within the less-conspicuous young, unfurling leaves of trees such as oaks and maples. The increased movement of water into the cell vacuoles of the developing leaves contributes to their rapid expansion and growth. As the young leaves of these trees expand and mature, the anthocyanins are replaced by the summer green of chlorophyll. Anthocyanin pigments add striking color not only throughout the autumn and spring to old and young leaves respectively, but also throughout summer to many flowers and fruits of our gardens.

Both anthocyanins and betalains help protect tender young leaves and older autumn leaves from both freezing temperatures and hot, dry ones. Both pigments localize to cell vacuoles, and their presence in vacuoles draws water in by osmosis. This ability to attract water serves to keep pigment-laden cells well hydrated even during droughts. On days of freezing temperatures, the presence of many pigment molecules dissolved in the water of each cell also has the beneficent effect of lowering the freezing point of water within the cells and preventing the formation of ice crystals that would rupture and kill the cell. The addition of pigment molecules to the water of cell vacuoles inhibits ice formation, just as the addition of salt molecules to water on roads prevents ice from forming there. So besides their brightening influence on a landscape, multitalented betalains and anthocyanins help ensure the safe passage of plant cells through stressful environmental conditions.

Like betalain pigments of beets and Swiss chard, anthocyanins are strong antioxidants that destroy the harmful free radicals that often form in our bodies and play havoc in our cells. The unpaired electrons of free radicals can oxidize and chemically alter any number of important cellular compounds. Antioxidants donate electrons to neutralize these free radicals before they can inflict damage on cells. A diet rich in fruits and vegetables embellished with red, blue, and orange pigments is a diet rich in antiaging antioxidants.

By absorbing ultraviolet light, these colorful antioxidants also act as sunscreens for plant cells. Moving green cabbage and broccoli seedlings that have been started indoors to the spring garden for transplanting triggers a color change from green to purple. Behind the glass of windows where they began germinating, the seedlings were protected from the ultraviolet rays of the sun. After a few days in the sun and exposure to its ultraviolet rays, however, the cells in the green leaves of the seedlings begin producing purple anthocyanin as a natural sunscreen.

HYPOTHESIZE: If anthocyanins have an ultraviolet-protective function in plant cells, then ultraviolet light from the sun should promote or induce the formation of anthocyanins in plant cells. Anthocyanins impart red to apples, peaches, and strawberries; purple to plums and grapes; and deep purple to the fruits of eggplants (fig. 8.7). Cells in the skins of these fruits have vacuoles loaded with anthocyanins. One can diminish or eliminate exposure of these pigmented fruit cells to sunlight by covering the ripening fruit of red apples or crabapples with a paper bag. How does anthocyanin pigment develop in ripening fruit that has been covered for several days? Will anthocyanin production begin anew after the bag has been removed and the fruit has been re-exposed to sunlight?

We can test the hypothesis that the ultraviolet spectrum of sunlight induces anthocyanin production in pigment cells of fruit by exposing different parts of a ripening apple or peach to different components of sunlight. While the colors of ripening apples are more green than red, wrap one large apple in clear, ultraviolet-proof polyvinyl chloride film (sold as Saran Wrap or Reynolds Wrap 914 in stores). Wrap another apple in clear, polyethylene film (sold as generic clear plastic bags in stores) that is not ultraviolet proof, and leave a third apple directly exposed to full sunlight. After the uncovered apple has

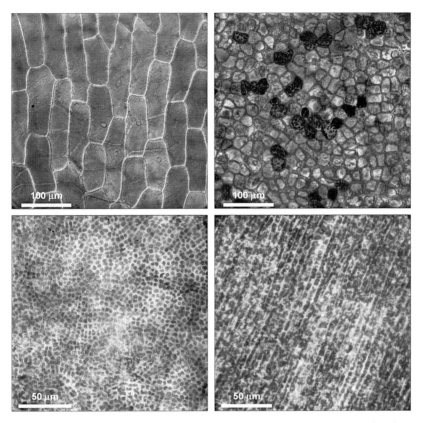

Figure 8.7 The pigment cells of fruit and vegetable skins contain anthocyanins within their vacuoles. *Top row, left to right:* red onion, blueberry. *Bottom row, left to right:* apple, eggplant.

attained its full red color, compare this natural color with the colors of the apples induced by (1) sunlight filtered through the film of polyethylene and (2) sunlight filtered through polyvinyl chloride film.

Do different exposures to sunlight influence the intensity of anthocyanin colors for the often pronouncedly red autumn leaves of a tree such as a maple, a dogwood, a sweetgum, or a blackgum? Do the more shaded interior leaves of these trees have less intense reds than their outermost leaves? Do trees growing on exposed ridges or open cityscapes have more intense fall colors than trees growing in deep, shaded ravines?

Plant Pigments That Can Be Used as Dyes

OBSERVE: Plants, their flowers, their fruits, and their roots come in a kaleidoscope of stunning colors. Collect a variety of vegetables, fruits, roots, bark, or flowers from plants whose appealing color you would like to impart to cotton, linen, or muslin fabric—or even boiled eggs (fig. 8.8). Chop the plant parts, add twice as much water as chopped plant material, bring to a boil and simmer for an hour, then strain and set aside this dye bath. To prevent these natural plant dyes from fading or washing out, fabrics or boiled eggs must be first exposed to a fixative, or mordant, bath. This fixation step ensures that the plant dye you subsequently add adheres and fixes to the fabric or egg so that color remains after washing. Therefore, after the dye bath has been prepared, place the damp fabric of your choice in a cold fixative or mordant bath (vinegar bath = 1 part vinegar to

Figure 8.8 So many different plants and parts of plants can be used as sources for natural dyes. Onion skins (*lower left*) impart an orange color. Purple grapes (*upper left*) offer a blue-purple dye. In addition to their appealing fragrance, the boiled leaves of basil (*upper right*) confer a purplish-gray color. The red fruit of sumac (*lower right*) provides a soft red color.

4 parts water). The best fixative for dyeing with berries, however, is prepared by dissolving half a cup of table salt (NaCl) in eight cups of cold water. Salt solutions and the acidity of distilled white vinegar act to adjust the charges on the fibers so that natural dyes bind more readily to textile fibers. Simmer the material in this hot fixative bath for an hour if you are dyeing cloth. Then remove the fabric, rinse it with cool water, and wring it out thoroughly. Place the wet fabric in the dye bath and simmer until the desired color is attained. (Remember that once it is dry, your fabric will be lighter than its wet version.) Remove the fabric from the dye bath with rubber gloves and wash it in cold water. Wring the fabric thoroughly and hang it to dry. Launder naturally dyed fabrics in cold water and separate them from other laundry.

When dyeing boiled eggs, submerge boiled eggs in a cold dye/fixative solution prepared by adding one tablespoon of vinegar to every cup of the natural dye bath; leave the eggs and dye/fixative solution refrigerated until the egg color seems intense enough. Carefully dry the colored egg and enhance its color by polishing it with a little vegetable oil.

HYPOTHESIZE: Can you easily predict what color a dye mixture of a particular plant part will impart to a fabric or a boiled egg? Weeds (such as dandelions or plantains), vegetables (such as beets or onions), and shrubs (such as sumac or dogwood) can each be sources of a range of earthy colors, depending on the part of the plant chosen as a source of dye. Try this natural dyeing process with boiled eggs to see how time, temperature, plant part, or vinegar concentration influences how vibrant, how dark, and even how unexpected certain natural dyes can be. After the dyed eggs have dried, see how polishing each with a little oil can brighten up an already colorful egg. Which of the colorful plant pigments—betalains, carotenoids, or anthocyanins—give the most intense colors?

9

PLANT ODORS AND OILS

The enticing odors of herbs—parsley, sage, rosemary, and thyme—emanate from tiny oil droplets found inside special cells of their leaves, flowers, and stems (fig. 9.2). While culinary herbs such as parsley and thyme add flavor to our foods, and other herbs such as catnip impart happiness to our cats, the oils of herbs have protected plants for millennia from the jaws of insects, long before humans and cats discovered the appeal of herbs. These oils are included in the category of plant chemicals referred to as secondary metabolites—chemicals that influence a plant's interaction with its environment but that are not essential for the immediate survival and reproduction of the plants that produce them. The oils might not be toxic to insects, but the odors of these oils can be repellent and repulsive to the

Figure 9.1 A mouse and a toad peer out from the herb garden. Scattered among the catnip, parsley, sage, rosemary, and thyme are insect parasites, herbivores, and pollinators. The soldier beetle (*upper left*) begins life as a fast-moving predatory larva in the soil but spends its adult days pollinating flowers. The tachinid fly and chickweed geometer moth (*lower left*) are pollinators of garden flowers, while their larvae have very different roles in the garden. The larva of the fly is a parasite of squash bugs, stink bugs, and leaf-footed bugs; the larva of the moth feeds on a common garden weed called chickweed. An adult stink bug (*lower right*) feeds on other insects, and the colorful caterpillar feeding on parsley leaves between the moth and the stink bug is the larva of the black swallowtail butterfly. The caterpillar has been alarmed by the toad and has extended bright orange horns on top of its head that emit an unpleasant, repellent odor.

antennae of many insects, even those that don't munch on plants. While oils and other secondary metabolites produced by plants can give some insects severe indigestion, many insects have found ways to circumvent these disagreeable encounters; and some have even developed a fondness for the flavors these oils and chemicals impart to leaves and stems.

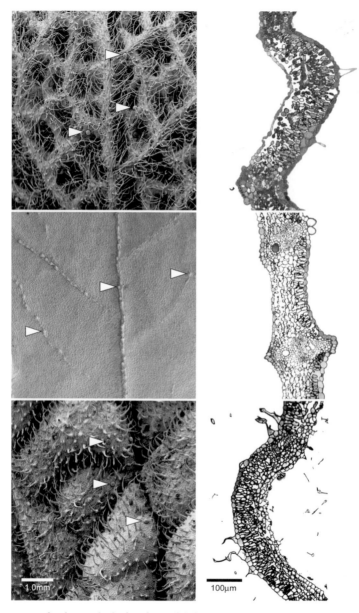

Figure 9.2 Left column: The leaf surfaces of different herbs have characteristic landscapes with distinctly shaped cells (*arrowheads*) that produce their distinctive odors. *Top to bottom:* Sage, parsley, and catnip. *Right column:* Sections through these same leaves reveal some of the stomata, hairs (trichomes), and glands of leaf surfaces that overlie the typical arrangement of cells nestled between the upper and lower monolayer of cells (epidermis) covering each leaf.

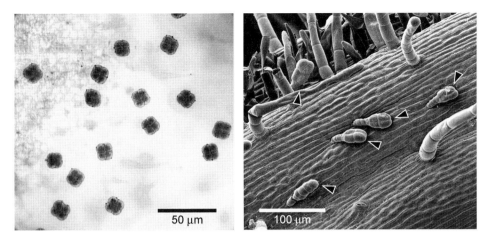

Figure 9.3 The epidermal surfaces of tomato (*left*, viewed with the light microscope) and birdhouse gourd (*right*, viewed with the scanning electron microscope) leaves and stems are covered with a fuzz of trichomes. On tomato leaves, the four-cell trichomes responsible for the peculiar odor of the tomato stand out like dark lollipops. On the tendril (a modified leaf) of a gourd are similar four-cell trichomes (*arrowheads*) scattered among taller, spindly trichomes.

The distinctive oils of plants reside in special glandular cells on their surfaces. The unmistakable fragrance of tomato plants comes from the numerous glandular trichomes (*tricho* = hair) that cover their leaves and stems like fine fuzz. Feel and smell the surface of birdhouse gourd leaves; the microscope reveals a fine velvety fuzz covering the surface of gourd leaves and stems (fig. 9.3). The fuzz consists of glandular trichome cells that stand out as distinctly shaped cells among the sea of uniformly shaped epidermal cells of the leaves. These glandular cells not only produce odors that repel some insects but also present a physical challenge for many tiny insects trying to navigate through these forests of fuzz.

The unmistakable odors of such plants as tomatoes and gourds that are distasteful to some insects, however, can be delectable to others. Three plant families found in gardens—cabbage, squash, carrot—are especially noteworthy for having chemicals only some insects find to be alluring and tasty (appendix A).

The pungent mustard odors of plants in the cabbage family such as broccoli, mustards, collards, kale, and turnips attract certain aphids, flea beetles, diamondback moths, and cabbage butterflies. Glucosinolates are the secondary metabolites that impart these mustard odors. These chemicals are potent anticancer agents in our own diets and are responsible for the health benefits of eating these popular vegetables, even though many insects, pathogenic fungi, and certain plants find glucosinolates and their derivatives to be toxic. Mustard odors represent chemicals with diverse talents and great versatility in the food web of a garden.

Plants of the carrot family such as dill, parsley, parsnips, carrots, and fennel likewise produce chemicals called furanocoumarins that are toxic repellents to some insects. But these same chemicals are rapidly converted to harmless forms by caterpillars of certain moths and swallowtail butterflies. For these mother butterflies and moths, as for other insect mothers, the choice of future plant home for their larvae is based on the peculiar odors of the plants that have nurtured and supported each of them for countless generations.

Members of the squash family (Cucurbitaceae) such as pumpkins, cucumbers, zucchini, and gourds contain bitter chemicals called cucurbitacins that repel many creatures; however, the chemicals actually attract certain insects that make their living feeding on members of this plant family. Among these insects that consume bitter cucurbitacins with impunity are cucumber beetles, squash bugs, and the caterpillar of a handsome moth called the squash borer.

Even though plants cannot run from danger, they can handily defend themselves when faced with challenges from insect jaws or hostile microbes such as fungi. Most plants have a variety of ways to repel unwanted visits from creatures that chew their leaves or suck their sap. Those pests that persist in spite of their initial encounters with plant odors or textures face other plant defenses. Plants respond to bites of insects and invasions of microbes by producing a variety of chemicals—some that discourage attacks from microbes and insects, giving some pests indigestion or proving lethal to others.

Plants That Send Warning Odors to Fellow Plants

The secondary metabolites that not only mediate interactions among plants but also defend plants from attacks by insects and other herbivores, fungi, and bacteria are estimated to number at least 200,000. Cells of many plants continually produce many of these chemicals such as glucosinolates and cucurbitacins to discourage insects from grazing on their leaves (appendix A). Tannins are also common compounds found in leaves of plants such as spinach (fig. 2.18) and oaks that can disrupt the digestion of plant-feeding insects by binding to enzymes and other proteins in insect guts. Rotenone, a naturally occurring insecticide produced by members of the pea family (Fabaceae), is a well-known secondary metabolite. These compounds that are constitutively produced by plants offer a first line of defense.

Phytoalexins (*phyto* = plant; *alexin* = defend) are another category of secondary metabolites that, by contrast, are produced only when bacteria or fungal pathogens invade plant cells. Secondary metabolites produced at one wound site on a plant prepare the plant for possible attacks elsewhere by sending chemical signals throughout the plant that not only increase its resistance to additional attacks but also forewarn neighboring plants of impending threats.

The plant being attacked releases a variety of chemicals generally referred to as volatile organic compounds, or VOCs for short. Chief among these are the simple organic compounds of methyl salicylate (produced from the plant hormone salicylic acid) and jasmonic acid. Another plant hormone, ethylene, also acts in concert with these many volatile compounds whenever plants face attacks from microbes or insects. These chemical signals waft through the air and are picked up not only by other plants as alarm signals but also as "cries for help" to recruit nearby insect predators and parasites that can eliminate whatever insects are devouring their leaves and that can help the plants in their distress. Along these lines, VOCs induce nearby plants to ramp up their own defenses against hungry insects. In response to the VOCs, plants commence producing a variety of new chemicals that provide an even more thorough layer of preemptive defense before the pests have even arrived.

OBSERVE: If plants can communicate "danger" to fellow plants by emitting volatile chemicals, do nearby plants respond by reinforcing and bolstering their chemical defenses against hungry insects? Can insects be discouraged from feeding on plants that have been forewarned of an impending insect attack? Leaves of vegetables such as broccoli, tomato, or cabbage are often chewed and shredded by caterpillars. If you spot a few caterpillars feeding on a few of these particular leaves, see if you can discourage—or even halt—additional feeding by these caterpillars. Simulate a traumatic insect attack on a single one of these same plants by shredding several of its leaves. Then compare over the next few days what happens to caterpillars that you have placed on nearby fellow plants (within a radius of five feet) with the fate of caterpillars that you have relocated to distant fellow plants (greater than twenty feet away). Do these leaf-chewing insects continue feeding, or do they disappear? Do some caterpillars grow faster than others? Does the damage inflicted on one vegetable species alert plant defenses not only in other members of that species but also in other nearby vegetable species?

As an additional preventive measure, you might also place a tiny vial of fragrant, volatile methyl salicylate (oil of wintergreen) among vegetables that are popular with plant-feeding insects. Does this VOC perform as predicted and repel insects from feeding on the vegetables? Does methyl salicylate halt, reduce, or have no influence on damage to vegetables once hungry insects have discovered them?

HYPOTHESIZE: If the natural plant hormone salicylic acid induces the production of defensive compounds in the plant that ward off attacks by insects, fungi, and bacteria, then could the related chemical acetylsalicylic acid, found in household aspirin, mimic this effect of the plant hormone (fig. 9.4)?

For thousands of years before aspirin became available in pill

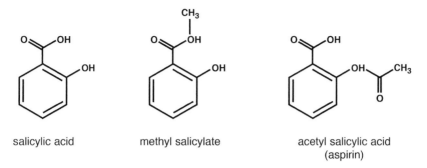

salicylic acid   methyl salicylate   acetyl salicylic acid
(aspirin)

Figure 9.4 The chemical structures of the natural plant hormone salicylic acid, the volatile derivative produced from this hormone known as methyl salicylate, and acetylsalicylic acid of aspirin are similar; all three of these simple compounds may function in defending plants from insect and fungal attacks.

form, people in both the Old World and the New had discovered that the inner bark of willow trees has the remarkable ability to relieve pain, inflammation, and fever. Not until the 1800s, however, was the single chemical responsible for these multiple medicinal benefits actually isolated from willow bark. The active substance was named salicylic acid, after its first botanical source—willow trees of the genus *Salix*.

Salicylic acid induces the production of other compounds in plant cells such as volatile methyl salicylate that function in defending the plant against attacks by foreign intruders—both small and large, both microbial and insect. The action of salicylic acid is referred to as conveying systemic acquired resistance (SAR) to a plant threatened by disease or insect damage. This hormone's action is the equivalent of triggering an immune response in plants. Can salicylic acid's chemical relative aspirin substitute for this plant hormone in its role of inducing SAR, in preventing attacks before threats of pests arise in the garden? Can aspirin act as a preemptive, preventive measure to protect vegetables in your garden that are usually vulnerable to attacks from pests? Dissolve one uncoated 325-milligram aspirin tablet in a gallon of water and two tablespoons of dishwashing soap along with a couple of drops of vegetable oil. The addition of soap and oil helps the spray adhere to the waxy surface of leaves. Try applying as a spray every two weeks to the leaves of plants in one row and then applying the same spray containing soap and vegetable oil but without the dissolved aspirin to another row of the same vegetable.

Plant Odors as Both Repellents and Enticing Fragrances

OBSERVE: Try observing with your nose. Go about the garden rubbing the leaves of plants with your fingertips and then sniffing your fingers to note the distinctive odors of so many of the fruits and vegetables of the garden. The distinctive odors of plants conveyed by their numerous secondary metabolites will become familiar as you associate with them during the year.

As they are picked up by the antennae of insects, odors associated with plants can either entice, repel, or be ignored. What proves to be repulsive to some plant-chewing insects often is also repulsive even to insects such as blood-sucking mosquitoes that have neither the desire nor the jaws to feed on plant leaves. Plant chemicals whose odors appeal to our noses but are repulsive to mosquito antennae would be ideal fragrances to use as insect repellents. Such chemicals may be discovered among the panoply of odors that we observe in a typical garden.

To extract and enhance the natural fragrances of herbs and flowers from the garden, a vegetable fat such as margarine is first used to absorb the volatile odors emitted by fragrant leaves or flowers. Most of the scents of plants are derived from their oils; and oils are more readily dissolved in vegetable fats and alcohol than they are in water.

Spread the margarine in a uniform layer about one-half centimeter or a quarter of an inch thick over a glass surface of a shallow dish that can be subsequently covered by a glass plate. Then gently lay the fragrant leaves and flowers on top of the vegetable fat. Without squashing the plant parts, place the cover over this arrangement and seal the plant tissues between the top and bottom halves of a large petri dish or inside a large shallow glass bowl (such as a pie plate) that can be tightly covered by a glass sheet or plate. Leave the dishes and plant parts undisturbed and out of the sunlight for two days at room temperature. At the end of this time, remove the plant parts and replace them with fresh plants. Repeat this procedure about four times to saturate the vegetable fat with a particular plant odor. At the end of the procedure, scrape the vegetable fat into a wide-mouthed bottle that can be securely sealed. Next, add about an equal volume of 95 percent ethanol or vodka. Repeatedly and vigorously blend and shake this combination. The plant odors will partition to the ethanol and can be separated from the vegetable fat by freezing the mixture. The fat will solidify, and the ethanol that now contains the plant extract can be poured off. Do any of these natural fragrances that appeal to our human senses actually prove offensive to insect senses?

HYPOTHESIZE: On days when female mosquitoes are biting well, see for yourself if any or all of the oils from six members of the mint family—catnip, basil, peppermint, sage, rosemary, and thyme—are as repellent to mosquitoes as catnip is enticing to cats (fig. 9.5). Begin by vigorously rubbing one arm from wrist to shoulder with catnip leaves or with the leaves of one of the five other fragrant herbs. Also rub one arm just as vigorously with *Coleus* leaves. Both *Coleus* and the six garden herbs are easy to grow, and they are all related. However, although *Coleus* is also a member of the mint family, it is the one member of the family that lacks the aromatic odors of the other six herbs. Count the number of mosquito bites that each arm receives. Is the odor of any or all of these herbs potent enough to repel mosquitoes from one arm, or even both?

Companion Gardening: Cooperation and Competition in the Garden

For centuries gardeners and farmers have observed that certain plants thrive in the company of specific plants while they languish

Figure 9.5 Herbs, such as those in the mint family, may have enticing odors for some creatures but repellent odors for others.

in the company of other plants. Gardeners practice companion planting to bring out the best attributes of their fruits and vegetables. The well-being of certain plants seems to be linked to the well-being of these other plants. They somehow complement each other's growth and prosperity; they communicate their mutual benevolence through the medium of air aboveground as well as through the medium of soil belowground (fig. 9.6). Our studies of this plant communication are still in their infancy; we are in the early stages of understanding the languages of plants.

Much mystique and mystery still enshrouds the practice of companion gardening. Solving this mystery will require knowledge of how plants exchange information and exactly what they exchange with one another. In the last three decades, scientists who have been eavesdropping on plant conversations aboveground have discovered that plants converse with one another by releasing specific chemicals

Figure 9.6 Cabbage (*left*) and sage (*right*) make good neighbors in the garden.

when hungry insects are chomping on their leaves. These signals from plants may have dual functions, some for the benefit—but in some instances for the disadvantage—of their plant neighbors. As we learn to eavesdrop on the communications among plants, we are deciphering a chemical language and dispelling some of the mysterious affinities and antagonisms of plant lives.

Scientists and gardeners know even less about what transpires underground between plants. But what scientists have found is that roots of plants have the ability to distinguish self from nonself. This ability is tantamount to a plant having an immune system, for our own immune systems enable each of us to ward off invaders such as microbes and other foreign objects. Our immune cells have the remarkable ability to recognize the difference between our own cells and all other cells. Plants also can mount a phytoalexin defense against foreign microbial invaders; and with allelopathic chemicals, they can inhibit the growth of roots from other plants.

In discussing how weeds deal with some of their competitors

(chapter 7), the phenomenon of allelopathy was introduced to demonstrate how certain chemicals—such as caffeine—secreted by the roots of coffee and tea plants can inhibit the germination of seeds from other plants, such as those of turnips. Black walnut is a well-studied example of a tree whose roots also secrete an allelopathic chemical that discourages the germination and growth of many other plant species. Like caffeine, the chemical secreted by walnut trees, juglone, is a simple organic compound. Tomatoes, potatoes, peppers and corn remain stunted when planted near walnut trees, but other plants such as dogwood, crocus, serviceberry, and daylilies seem immune to the inhibitory influence of juglone. Black locust, cottonwood, sycamore, tree of heaven, sassafras, sugar maple, and sumac are among other trees known to suppress the growth of certain plant species that try to grow beneath them. Wildflowers such as goldenrod as well as grasses such as fescue and bluegrass form such dense stands of plants that few other plant species can establish their rootholds among them. The success of invasive plant species has been attributed to their secreting allelochemicals that are particularly effective at stifling the growth of other plants. The chemistry responsible for the inhibitory allelopathic actions of many of these plants is currently being explored, and the chemicals produced by these plants are turning out to be relatively simple natural compounds that are being considered as environmentally safe alternatives to harmful synthetic herbicides (appendix A).

Specific allelopathic interactions among vegetables and flowers in our gardens may account for many of the so-far inexplicable influences of one plant on another's well-being. Cabbage, tomatoes, asparagus, cucumbers, sunflowers, and soybeans are all cultivated vegetables reputed to have allelopathic properties, but confirmation of the existence of these properties awaits closer observations in the garden. While these particular vegetables suppress the growth of some plants in their neighborhoods, however, they apparently promote growth of other vegetables (fig. 9.7).

OBSERVE: The architecture of roots is a reflection of how plants interact belowground. Plants sharing the same soil—its nutrients and its water—can either compete for the same resources or share their resources and coexist. They coexist best when they partition the underground resources by sending forth roots in different directions, extending roots to different depths, and spreading their roots sometimes short distances and sometimes sprawling long distances horizontally, at various depths beneath the surface. This is true for whatever combination of plants share the soil—weeds with weeds, vegetables with vegetables, weeds with vegetables. As you remove weeds from the garden, observe the differences in their root architectures. Which weeds have major taproots that vertically dive deep into the soil? Which weeds have multiple roots that spread horizontally?

Plant a single vegetable plant in a gallon pot (pot 1) and a single weed in another gallon pot (pot 2). Plant two of the same vegetable plant in the same gallon pot (pot 3) and two of the same weed in another gallon pot (pot 4). Plant the garden vegetable in the same pot with the weed you have chosen (pot 5), and finally plant the garden vegetable with another weed or vegetable of your choice (pot 6). As hinted by figure 9.7, the number of different combinations from which to choose is vast.

After growing the plants for a couple of months in the same soil and in the same aboveground environment, remove the roots from the pots and note how the root architecture of a plant is influenced by its neighbor. What you observe about these soil companions may shine some light on how roots interact belowground and process information about the presence of nutrients and neighboring roots (fig. 9.8).

Deciphering what information plants convey in the language spoken by their leaves and roots may allow us to understand better why tomatoes love carrots, sage loves cabbage, but why tomatoes avoid cabbage and cucumbers avoid sage.

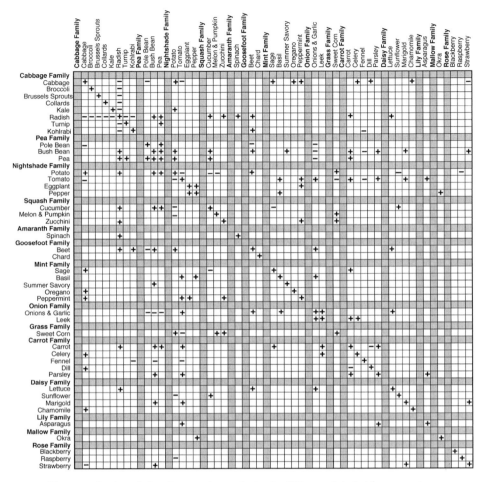

Figure 9.7 Our knowledge of companion gardening is still incomplete; but from experiences of many of our gardening ancestors, we can begin systematically recording the positive and negative influences of vegetables on each other. Of course, the interaction between a pair of different vegetables may be neither obviously positive nor obviously negative. Most of the information in this table was gleaned from the pages of the gardening classic, *Carrots Love Tomatoes: Secrets of Companion Planting for Successful Gardening*, by Louise Riotte. Future discoveries in our gardens will refine this table and fill in its gaps.

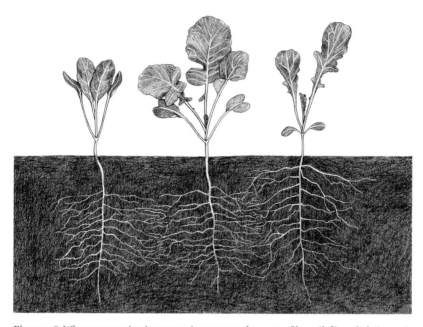

Figure 9.8 What communication transpires among the roots of beet (*left*), radish (*center*), and lettuce (*right*)? For some as yet inexplicable reason(s), these three vegetables all thrive in each other's company.

HYPOTHESIZE: Carrots and tomatoes are supposedly good companions. Based on the information that cabbage and tomatoes, however, seem to be incompatible, would you predict that cabbage and carrots would grow well together?

One member of its own carrot family—dill—seems inhospitable to carrots themselves, even though both carrots and tomatoes seem to thrive in the company of another member of that family, parsley. How would you expect dill to influence the growth of tomatoes?

HYPOTHESIZE: All members of the cabbage family (Brassicaceae) contain simple chemical compounds called glucosinolates that are precursors of mustard oil—a chemical whose pungent flavor is exceptionally enticing to small leaf-chewing flea beetles, cabbage butterflies, and diamondback moths. The most familiar members of

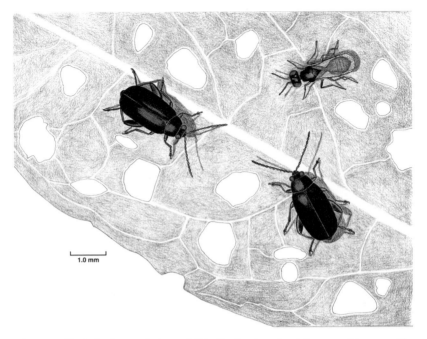

Figure 9.9 The holes in arugula leaves left by flea beetles attract tiny parasitic wasps. The larvae of these wasps feed on flea beetles and help keep flea beetle populations in check.

the cabbage family in American gardens are crops such as cabbage, collards, broccoli, cauliflower, kohlrabi, Brussels sprouts, and kale; these are all varieties of one species, *Brassica oleracea*. Other members of the cabbage family, such as Chinese cabbage, mustard greens, bok choy, tatsoi, radishes, turnips, and arugula, belong to different species in the genus *Brassica*—or even a different genus in the case of arugula. These other members of the family also have the distinction of being especially attractive to flea beetles—so attractive that their leaves are often ravaged and stripped by hordes of these little herbivores (fig. 9.9).

An effective method of protecting certain vegetables from the jaws of a particular pest is to offer the pest an alternative food that it finds even more appealing. Mustard greens are one of the most appealing members of the cabbage family to flea beetles. Because of their appeal, mustard greens have been employed as a "trap crop" to exploit the inordinate fondness of flea beetles for the odor of glucosinolates and mustard oils. The member of the cabbage family whose odor is most enticing to the flea beetles serves as a sacrificial offering to the flea beetles so that other vegetables in the cabbage family may survive with few, if any, flea beetle bites.

Seed catalogs now offer a wide selection of greens from the cabbage family in addition to such familiar and conventional crops as broccoli, kale, and cabbage. When "trap crop" seeds are sown, see which varieties are most effective in diverting the attention of flea beetles from feeding on less well-scented vegetables. Plant some members of the cabbage family without a nearby "trap crop" and compare the damage inflicted by flea beetles. Can you establish if chemical differences in mustard oil content and/or physical difference in leaf surfaces (waxy or hairy; tender or tough) contribute to the flea beetle response?

10

FELLOW GARDENERS: OTHER CREATURES WHO SHARE OUR GARDENS

Any successful gardening project actually represents collaborations with countless creatures, some with no legs, some with four legs, some with six and some with more. Earthworms, birds, insects, spiders, fungi, millipedes, innumerable bacteria, and many others share the garden with us and with the vegetables, aerating the soil, recycling its nutrients, and mixing its organic and mineral components. Together these creatures have developed ways to coexist so that the fruits and vegetables of the garden can be shared without some creatures harvesting entire crops of certain fruits and vegetables. Together these creatures form a food web where energy and nutrients are constantly exchanged while maintaining a balance in the numbers and activities of all members of the food web.

Figure 10.1 In the shade of the tomato plant, a mouse and a toad watch an earthworm drag a decaying leaf into its burrow. A predatory ground beetle darts past the mouse. On the left, a midge hovers in the sky above the toad after leaving its larval home in the garden soil. Between the midge and the toad, a hornworm caterpillar rests on a tomato stem, and a robber fly surveys the area for prey from its perch on a tomato leaf.

Also remember that plants are very effective in actively recruiting other creatures to come to their defense. Plants respond to bites of insects and invasions of microbes by producing a variety of chemicals—some discourage or even ward off attacks from microbes and insects by giving pests indigestion, but some of the chemicals drift through the air to attract predators and parasites of the insect pests, even recruiting these insects to help the plants in their distress. Other natural chemicals are broadcast on air currents to neighboring plants, acting as warning signals to these nearby plants that pests are out and about and triggering their own production of new chemical defenses before the pests have even arrived on these neighboring leaves.

When garden habitats are designed to provide inviting homes for helpful insects, microbes, and other gardening partners, these fellow gardeners far outnumber those insects and other creatures that we consider pests. Pests have a difficult time establishing a foothold and mostly go unnoticed when gardens are home to an abundance of predators and parasites both aboveground and belowground as well as to numerous recyclers of the soil. As they go about their business, recyclers enrich the soil, providing nutrients and structuring the soil habitat for roots in ways that improve the vigor of the plants. Thriving and productive garden habitats attract a diversity of creatures—large, small, and microscopic; their healthy soils nourish healthy plants and fortify plant defenses against pests and pathogens.

Microbial Partners of Plants

Rhizobial Bacteria

OBSERVE: Members of the pea family that grow in our gardens (peas, beans, clovers—collectively known as legumes) have a special alliance with nitrogen-fixing bacteria that have the ability to take nitrogen gas from the air, converting it to forms of nitrogen that plants can use. Bacteria are the only creatures on Earth with this nitrogen-fixing ability, and these rhizobial bacteria that inhabit nodules on roots of peas and beans represent a special group of these nitrogen-fixing bacteria. If the soil of your garden has been abused like so many soils that are treated like "dirt," you might want to add what is called a rhizobial (*rhizo* = root; *bios* = life) inoculant to the bean and pea seeds that you plant. These inoculants are found in all seed catalogs and garden stores. The addition of the inoculant will ensure that there are sufficient numbers of bacteria to provide essential nitrogen for the robust growth of the beans and peas—and often enough nitrogen for neighboring vegetables as well.

Each summer I plant three crops of green beans in my central Illinois garden—the first crop is planted in early May, the second crop in early July, and the third in early September. After harvest of each crop, I pull up the bean plants and add them as mulch to another section of the garden. But as I pull up each old bean plant, I admire the cooperation between the bean roots and the rhizobial bacteria that is responsible for the formation of spherical nodules that coat the roots. Such teamwork provides essential nitrogen for the plant and a stable habitat and energy resources for the bacteria. Within each of these nodules, millions of rhizobial bacteria carry out the job of producing a form of nitrogen—ammonia—that plants can readily use. Although plants are surrounded by air that is roughly three-fourths dinitrogen gas, plants are totally reliant on bacteria to convert what for them is inaccessible dinitrogen to accessible ammonia. As these nitrogen-fixing bacteria move from the nearby soil, they first gain a foothold on fine root hairs on the root's surface, from which they then move deeper into the root, becoming encapsulated within individual root cells and inducing these root cells to divide and expand. The cells of the root assume special forms as each cell grows to accommodate and enshroud thousands of bacteria within their many membrane-bound vesicles (fig. 10.2).

A look inside a single root nodule reveals root cells filled with bacteria that produce an elaborate enzyme called dinitrogenase. This particular enzyme works best at breaking the strong bonds holding the two nitrogen atoms of dinitrogen together when there is no oxygen to obstruct its activity. To minimize exposure of this finicky enzyme to oxygen, the plant cells have made a special effort to come to the aid of their rhizobial bacteria and their hardworking dinitrogenase. As part of the collaborative symbiotic association of plant cells and bacterial cells, the root cells of legume plants produce a red iron-containing protein related to our human protein hemoglobin (called leghemoglobin) that avidly binds oxygen. By binding up oxygen in the vicinity of the root nodule, leghemoglobin improves the efficiency of the bacterial dinitrogenase as it goes about fixing nitrogen.

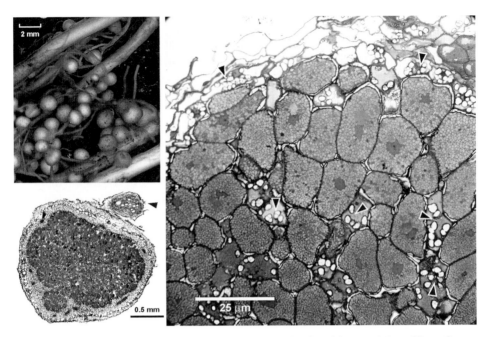

Figure 10.2 *Top left:* The roots of green beans are covered with nodules containing millions of nitrogen-fixing bacteria. *Bottom left:* A section through a small bean root (*arrowhead*) and its attached nodule shows an inside view of one entire root nodule and the dark root cells in which rhizobial bacteria dwell. *Right:* Each nodule contains many root cells that have each grown to accommodate the countless bacteria that have moved into their numerous membrane-bound chambers. The spaces between the approximately fifty stained root cells are occupied by numerous white starch granules (*arrowheads*)—a source of energy for the bacteria. (The outer edge of the nodule is on the top side of the image.)

About one hundred years ago, scientists discovered their own method for fixing nitrogen—that is, generating nitrogen fertilizer, or ammonia, from dinitrogen gas. They also discovered, however, that breaking the strong bond holding the two atoms of nitrogen in each molecule of dinitrogen gas (N_2) and converting it to ammonia (NH_3) requires vast amounts of energy, using between 3 and 5 percent of the world's natural gas production. Now, each year, more than a hundred million tons of energy-demanding synthetic fertilizer is applied on agricultural fields worldwide, usually in the form of ammonium nitrate (NH_4NO_3).

Plants best utilize small pulses of this nitrogen fertilizer, but large doses of fertilizer are most often applied to a field for reasons of convenience. Too much nitrogen fertilizer increases the acidity of the soil, and when applied on poorly drained soils can be lost as two forms of nitrogen gas called nitric oxide (NO) and nitrous oxide (N_2O). The latter gas is one of the most potent of the greenhouse gases that are known to absorb heat from the sun—almost three hundred times more potent than the best-known greenhouse gas, carbon dioxide. Since synthetic fertilizer unfortunately is applied in far larger doses than plants can use at any given time, these doses turn out to present a toxic shock to the many soil creatures that live in the path of fertilizer application. Also, since plants cannot use all the applied nitrogen fertilizer at once, much of it often runs off the soil or is leached out of the soil by rain. The leaching of negatively charged nitrates drags positively charged nutrients such as calcium, magnesium, and iron along with the nitrates. As excess nitrates flow into rivers and on to the sea, they promote the excessive growth of algae and aquatic weeds whose eventual death and microbial decomposition exhausts the oxygen supply for other aquatic organisms. For all these multiple reasons, the widespread use of synthetic fertilizers takes a costly toll on the environment.

The rhizobial bacteria of soil that inhabit the root nodule cells of plants are only a few of the countless bacteria of the garden soil that provide nutrients and safeguard the health of plants. After all, an estimated 10 trillion (10^{13}) bacteria live in the top six inches of a square meter of well-tended garden soil. The extremely diverse bacteria of the soil can carry out chemical transformations that no other creatures on Earth are capable of achieving. Among these innumerable bacteria are some that do not live in plant root nodules but that can still fix nitrogen as free-living bacteria. These and other bacteria of the soil can provide nutrients that otherwise would be inaccessible to plant roots; they can convert toxic chemicals to harmless, even useful substances. With their vast array of antibiotics, they can control the actions of harmful bacteria and fungi. Bacteria of the soil are master chemists that work essential untold wonders for the garden.

What has recently been discovered is that the plant hormones salicylic acid, jasmonic acid, and ethylene not only protect leaves and stems of plants from attacks of insects, bacteria, and fungi, but also control which bacteria—of all the innumerable and diverse bacteria in the surrounding soil—take up residence in and around the roots to help them with uptake of nutrients and with protection from pathogens and root eaters. Secretion of sugars from plant roots attracts bacteria of the soil. Bacteria congregate in great numbers around plant roots to partake of these gifts. However, plant roots are very selective about which bacteria move into their root tissues. Hormones secreted by roots inhibit the growth of some soil bacteria while encouraging the growth of other bacteria. Consequently, the assemblage of bacteria that dwell among the root cells of a plant are often very different from the assemblage of bacteria that reside in the surrounding soil. Plant roots are very selective about the microbes with which they keep company.

Mycorrhizae

We are beginning to appreciate the importance of the companionship that almost all plant roots—wild and domesticated—have developed with mycorrhizae (*myco* = fungus; *rhizae* = roots). Those fungi that are companions of trees form mushrooms aboveground and alliances with tree roots belowground (fig. 10.3). Their fungal filaments enshroud the roots of trees and penetrate between root cells, exchanging nutrients at the fungus-root interface but never penetrating the delicate cell membranes of root cells and harming these cells in any way. Fungi and trees are connected in networks where nutrients are shared and exchanged among multiple trees and fungi. Because these fungi form sheaths on outer surfaces of roots, they have been aptly named ectomycorrhizae (*ecto* = outside).

However, the mycorrhizae that keep company with garden

Figure 10.3 The fungal filaments that form ectomycorrhizal alliances with the roots of trees enshroud the root tips and form extensive underground networks for nutrient exchange among fungi and neighboring trees. The mycorrhizal fungus uses nutrients it obtains from the trees with which it is allied to produce its mushrooms aboveground and provides its green partners with water, mineral nutrients, and protection from soil pathogens.

vegetables and just about all the thousands of species of herbaceous plants have developed an even more intimate relationship with their green partners. These fungi do not form mushrooms and are strictly subterranean. They have given up their independence to become partners with green plants. The filaments of this group of mycorrhizae not only penetrate between root cells but actually enter inside the cell walls of the root, interdigitating with the membrane surfaces—but never penetrating through—the membranes of these cells. These endomycorrhizae (*endo* = inside) are also known by the name vesicular-arbuscular mycorrhizae (VAM). Not only do they form spherical vesicles (*vesiculus* = little bladder) within root cells, but they adopt highly branched, treelike forms, or arbuscules (*arbor* = tree; *-culus* = little) within individual cells. By adopting these multiple-branched forms, these fungal filaments vastly increase the surface area over which the root cells and fungal cells can exchange nutrients, water, and apparently substances that help defend them from whatever enemies lurk in the underground (fig. 10.4).

The seeds of one plant family—actually the largest of all the flowering plant families, the Orchidaceae—will not sprout and grow without help from mycorrhizae. The long and thin seeds of the approximately 26,000 species of orchids are noteworthy for their small sizes, ranging in length from 0.05 millimeter (half the width of a human hair) to 6 millimeters. Seeds this small only have room to accommodate the embryo of the future plant and lack the nutritive endosperm tissue that forms at the time of pollination in other flowering plants (chapters 1 and 4). A good portion of most seeds is devoted to endosperm, the nutritive storage tissue that nurtures the embryo through its vulnerable early days of growth. Without endosperm, however, orchid embryos have come to rely on fungal partners to supply the nutrients for their early days until the young orchids gain their rootholds and commence photosynthesis.

Just as plant roots have a say about the bacteria with which they associate, roots also carefully control the fungi with which they establish relationships. A newly discovered class of hormones secreted by root cells is known to attract mycorrhizae of the soil. These hormones, however, were first discovered and named strigolactones for their role in stimulating the germination of a parasitic weed called *Striga* (*striga* = witch), or witchweed, that can be a major pest of crops in parts of Africa. A germinating *Striga* seed uses the hormone to locate roots of its green plant host, pushes through the host's root cells, and invades the root's nutrient channels of xylem and phloem.

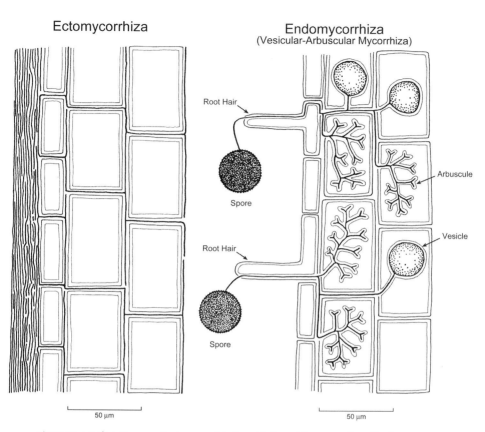

Ectomycorrhiza

Endomycorrhiza
(Vesicular-Arbuscular Mycorrhiza)

Root Hair

Spore

Root Hair

Spore

Arbuscule

Vesicle

50 μm

50 μm

Figure 10.4 Left: A diagram of ectomycorrhizal association with rectangular cells of tree roots shows that fungal filaments form a sheath on the surface of the root (*left side of image*), with individual filaments extending around and among root cells but never actually penetrating root cell walls or root cell membranes. The thinnest lines represent root cell membranes, and the thickest lines represent filaments of the mycorrhizal fungus. Lines of intermediate width correspond to the cellulose cell walls of root cells. *Right:* In this diagram of an endomycorrhizal association with the rectangular cells of herbaceous plant roots, the fungal filaments branch into treelike forms by penetrating the walls of root cells and interdigitating with—but not piercing—their cell membranes.

This parasitic partnership—unlike the mutually beneficial mycorrhizal partnership—benefits only the parasitic witchweed as it depletes the nutrient supply of its host. While these parasitic witchweeds have exploited the normal hormonal signaling of plant roots for their own purposes, the benefits that roots of green plants derive from attracting mycorrhizal fungi in the soil with strigolactone hormones far outweigh the harm derived when witchweed uses these same root hormones.

OBSERVE: How ubiquitous are mycorrhizal relationships between roots of green plants and soil fungi? To check for the presence of mycorrhizae on thin young roots of plants in the garden, a simple staining procedure has been developed. Chlorazol black E is a stain that labels the chitin of fungal cell walls but not the cellulose of plant cell walls. The cell walls of plant root cells remain unlabeled since they lack chitin. One part chlorazol black E solution (1 percent in water) is added to one part glycerin and one part lactic acid. The roots are left in this dark solution overnight or for several days and finally transferred to the clear solution containing one part glycerin and one part lactic acid on a microscope slide. When viewed with a microscope, the root tissue will appear clear, with only clear root cell walls and black-stained fungi visible (fig. 10.5).

HYPOTHESIZE: To judge the importance of partnerships between vegetables and mycorrhizae, see if mycorrhizal inoculants actually improve the lives of vegetables in the garden. Inoculants of mycorrhizae are commercially available from nurseries and seed companies. Try adding inoculant to half a row of one vegetable and leaving the other half of the row without inoculant. Measure one or more features of the vegetable that are clearly associated with robustness—for example, height of the plant, size of basal leaves, number of fruits, size of fruits. Do you observe clear differences in any of these features when you compare the vegetables grown in the presence of mycorrhizal inoculant with those grown in the absence of inoculant?

The observations that vegetables in the garden can have positive, negative, or indifferent influences on one another may have multiple explanations. One explanation assumes that mycorrhizae are involved in the interactions among vegetables. Specific vegetables and mycorrhizae might form hidden webs of connections underground

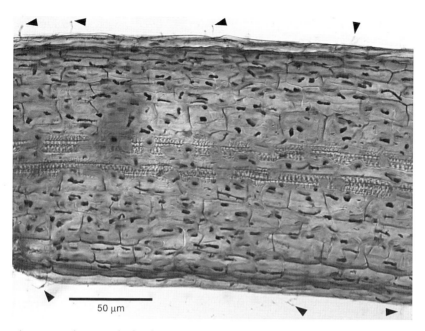

Figure 10.5 The network of endomycorrhizal fungi interdigitates among root cells of pepper plants and stands out as dark filaments, arbuscules, and vesicles after having its chitin-containing cell walls specifically stained with the biological stain chlorazol black E. Fungal filaments entering the pepper root from spores in the soil are marked with arrowheads.

that help account for why certain companions in the garden benefit from specific alliances. A failure of other vegetable companions to establish mycorrhizal networks would predict that these vegetables do not support one another or that they fail to thrive in one another's company.

Like bacteria, fungi are very diverse and multitalented, taking on a variety of tasks in the garden. Additionally, some fungi act as biocontrol agents, patrolling the soil as predators of other fungi, nematodes, and insects. Maintaining an inviting and alluring environment for all these diverse, beneficial fungi is the job of other fungi that serve as recyclers along with their supportive army of microbial and animal decomposers. These innumerable decomposers create a tantalizing underground habitat, and countless microbial and invertebrate allies of gardeners move in and occupy. According to the title of a gardening book by my friend Tony McGuigan, "Habitat it and they will come."

Decomposers and Recycling of Nutrients

Countless fungi, bacteria, and other microscopic creatures called microbes help plants by initiating the return of nutrients to the soil. These microbes recycle dead plant and animal matter and ensure that essential nutrients are always around for plant growth (fig. 10.6). Constantly taking vegetables and fruits from the garden along with all the nutrients they contain depletes the nutrients of the garden soil. We help return these lost nutrients when we fertilize gardens with manure, compost, and mulch, providing the recyclers and decomposers with dead plant and animal matter that is essential for their survival.

Assisting the microbes in their recycling endeavors are larger, more familiar recyclers of the soil such as earthworms. Earthworms help breakdown big chunks of plant litter as they chew, ingest, digest, and defecate, making it easier for the microbes to do their job. When everyone's job is complete, the recyclers have returned all the lost nutrients—in fact, more—to the garden soil that is again enriched and ready to host a new season of vegetables and fruits. Recyclers improve the chemistry of soil by adding to its nutrient content. By mixing the mineral particles of sand, silt, and clay with organic matter, the larger recyclers also transform solid, dense soil structure into loose aggregates that give soil a spongy structure with countless pores through which air, water, and roots can easily pass (fig. 5.8).

OBSERVE: Observing the microbial recyclers in your garden's soil— such creatures as bacteria, protozoa, and filaments of fungi that measure less than a millimeter in length or width—requires observing with a compound microscope; but the larger recyclers who assist the microbes and earthworms with decomposing and recycling can

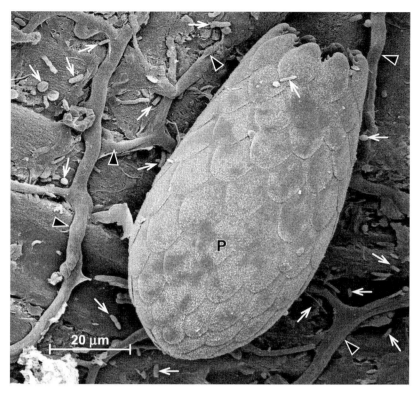

Figure 10.6 Among the decaying plant matter in a garden, microbial recyclers secrete enzymes that help break down this leaf surface. All the major microbial recyclers are represented in this image taken with a scanning electron microscope: the ornate shell of a protozoan (P); the filaments, or hyphae, of fungi (*arrowheads*); and representative bacteria of various sizes and shapes (*arrows*).

be easily collected with the help of a simple funnel, and just as easily viewed with a magnifying glass if a stereomicroscope is not available. Any funnel-shaped container can serve as a Berlese funnel (fig. 10.7). A half-gallon or gallon plastic bottle works quite well. These recyclers that measure at least one millimeter in length or width will include earthworms, snails, and the most abundant and diverse group of soil creatures, the arthropods. The latest estimate for the number of arthropod species worldwide is 1.4 million. All arthropods have jointed legs and include insects, mites, springtails, millipedes,

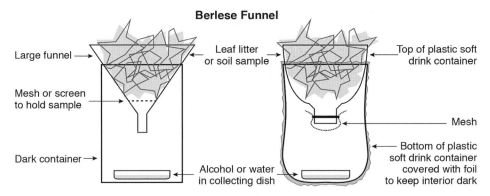

Figure 10.7 To avoid heat and light from above, dark-loving creatures inhabiting a sample of soil or leaf litter that is placed in a Berlese funnel will fall into the collecting dish below. Creatures can be collected and easily viewed on the white surface of a wet cloth or wet filter paper placed in the collecting dish beneath the funnel. The creatures can be returned to their habitat after they have been observed and appreciated.

and a variety of far less familiar but still important creatures found only in soils, such as the odd and peculiar proturans and pauropods (fig. 10.8). Collect the residents in a dish that is lined with wet filter paper or a wet paper towel, observe their forms and their habits, and then return them to the soil. Armies of these recyclers live in healthy garden soil with an abundance of organic matter. These recyclers may be small, but so many thousands live and work in every square meter of garden that they quickly decompose plant remains and return their nutrients to the soil. Considering that the arthropods alone in the top six inches of this square meter number around 150,000, expect to be amazed by their abundance and diversity.

HYPOTHESIZE: Compare the growth of particular vegetables growing with and without the addition of organic amendments to the soil. Such additions as compost, blood meal, bone meal, wood chips, horse manure, green grass clippings, shredded autumn leaves, a winter cover crop, and straw all contribute nutrients and organic matter to garden soil. Which organic additions to garden soil would

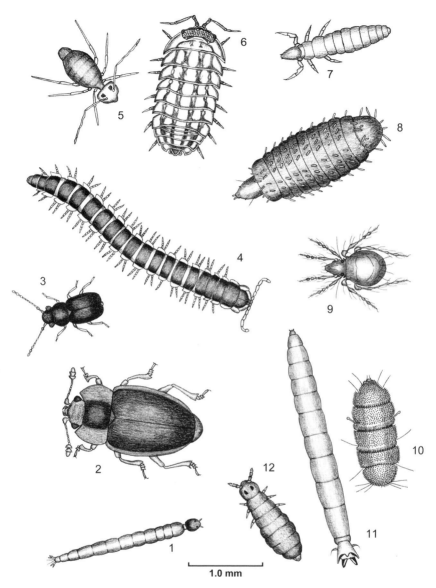

Figure 10.8 In addition to the invisible microbes (bacteria, protozoa, and fungi), many larger but still tiny creatures—best viewed under a magnifying lens—help out with the process of decomposing and returning nutrients to the soil to nurture a new generation of plants. Not only do they return nutrients to the soil, but they also mix organic matter with mineral particles to improve the physical environment for growth of roots. All creatures shown are arthropods collected with a Berlese funnel from garden soil that provides a welcoming habitat for recyclers and decomposers. These are only a few representatives of the great diversity of creatures you can find. *Clockwise from lower left corner:* (1) larva of a midge, (2) sap beetle, (3) featherwing beetle, (4) millipede, (5) globular springtail, (6) wood louse, (7) proturan, (8) larva of a soldier fly, (9) oribatid mite, (10) pauropod, (11) larva of a crane fly, (12) smooth springtail.

you predict best succeed in enhancing the growth of vegetables and which attract the greatest number and diversity of recyclers? Which amendments not only improve the chemical properties of a soil but also confer a sponginess to soil structure by the mixing of organic matter with the mineral particles of soil? Organic matter that remains after the thorough decay of plant litter is mixed by the actions of the larger recyclers, binding mineral particles of soil into small aggregates that are separated by pores, allowing the free passage of water, air, and roots into the spongy soil.

Attract recyclers to your garden by adding organic materials that provide ideal food and snug shelter for these essential members of nature's food web (fig. 10.8). But in order to carry out their mission of returning nutrients to the garden soil, recyclers need nutrients of their own to survive and multiply. As microbes work away, they must grow and multiply and must use whatever nutrients are available for their own growth and well-being. The one essential nutrient that is often in short supply is nitrogen (N), and nitrogen is an element needed for the manufacture of all proteins and all nucleic acids.

The relative level of nitrogen in cells and soil is expressed as a ratio of the nutrient nitrogen to the ever-present nutrient carbon (C). Microbial cells have a carbon:nitrogen (C:N) ratio of about 15:1 and must maintain this ratio of nutrients as they grow and multiply. When the material being recycled in the vegetable garden is relatively low in nitrogen and exceeds this ratio (C:N > 15:1), the microbes must rely on whatever nearby sources of nitrogen exist and take what they need for their own use. A local shortage of nitrogen results, which stifles the growth of nearby vegetables. Microbes that are recycling organic material with a C:N ratio greater than 15:1 actually deplete the nitrogen that would normally be used by vegetables of the garden. Once the C:N ratio drops below 15:1, however, the busy microbial recyclers begin adding nitrogen to the soil. Certain organic amendments to the garden actually suppress vegetable growth because their addition forces microbes to compete with vegetable roots for the limited supply of nitrogen provided by, for example, decomposing straw and wood chips, both of which are relatively low in nitrogen (C:N ~50:1 for straw and C:N ~500:1 for wood chips). Plants deprived of nitrogen are stunted, with leaves that are more yellow than green. Watch out for these telltale signs.

Plant a long row of one vegetable such as lettuce, beet, green bean, or spinach. Divide the row into six equal lengths. Do not add any organic amendment to one of the six equal stretches of the row, but add some of the following organic additions to the soil along each stretch of the remaining five lengths: wood chips, straw, fresh-cut green grass, tree leaves, bone meal, blood meal. Which addition do you predict will stimulate vegetable growth the most? Which addition will stimulate growth the least? Which addition will actually suppress plant growth?

Insects That Help as Predators, Parasites, and Pollinators

With their scents and colors, flowers of garden plants entice many animal visitors to spread their pollen. Some flowers attract predators such as wasps, certain beetles, and flies that not only spread their pollen from flower to flower but also act as allies in controlling certain unwanted visitors. Bees and butterflies are strictly pollinators, but wasps and flies play two key roles in the garden: they not only pollinate flowers, but also prey on many insect pests. Some wasps sting other insects and carry them to their nests as food on which their larvae feed and grow. Other wasps and some flies place their eggs on unsuspecting insect pests. When the eggs hatch, the wasp larvae and fly larvae dig beneath the skin of the pest. There they settle down as parasites feeding inside the body of their insect host. The host insect provides shelter and food for the parasites until they finish feeding and begin transforming to adult flies and wasps. Most parasitic insects (referred to as parasitoids) not only grow at the expense of their hosts but also eventually kill the hosts. Many of the wasps, beetles, and flies that visit flowers as adults often began life feeding on other insects—either as predators from the outside or as parasites from the inside. In gardens where these predaceous and parasitic insects thrive, plant-eating insects do not become a problem but rather can contribute to a harmonious balance of nature among the garden's many insect inhabitants.

Some beetles and some bugs—and even some caterpillars of butterflies and moths, such as the cabbage white butterfly and the diamondback moth—munch on the leaves of their favorite vegetables; in gardens where insects of all occupations are encouraged to live, these vegetable eaters may share a small part of the produce but will still leave plenty for others. In a garden where plant eaters and insect eaters live together in harmony, where pesticides are banned, there is an abundance of vegetables for insects and humans alike. A garden with a diversity of flowers and vegetables is an alluring habitat for the diverse community of creatures that naturally maintains harmony among its herbivores, its predators, and its parasites (fig. 10.9).

HYPOTHESIZE: By planting flowers—especially native flowers—to attract flies, wasps, beetles, bees, butterflies, and moths, not only do we attract insects to pollinate our crops, but some of these insects also prey on caterpillars and other insects that feed on vegetables (sometimes for their own meals but most often to provide food for their larvae, which feed as parasites or predators). We would predict that insect herbivores would have fewer opportunities to inflict damage in vegetable gardens that have neighboring flower gardens or that have flower rows interspersed with vegetable rows. Besides, the presence of flowers and small pockets of wildness enhance the beauty of any vegetable garden. Promoting the diversity of plants and flowers in a garden invariably promotes the diversity of not only its insect inhabitants but also its microbial and other animal life.

After the flowers have faded, the grasses have dried, and winter arrives, the remains of the summer garden offer a cozy refuge for those insects and their relatives that are such important gardening partners. At one or another stage in their development—egg, larva, nymph, pupa, adult—insects take a break from their warm-weather

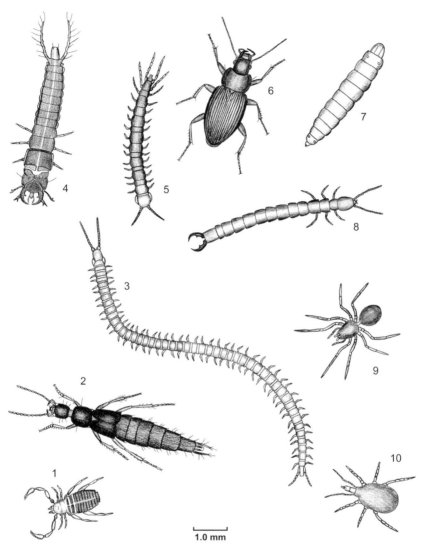

Figure 10.9 In addition to the more obvious and better-known toads, birds, ladybird beetles, and praying mantises that protect gardens from insect pests, countless small creatures live in the garden soil and help maintain a harmonious balance of nature in their roles as predators. These are just a few of the typical arthropod predators that can be found in a soil where gardeners work in partnership with nature. *Clockwise from lower left corner:* (1) pseudoscorpion, (2) rove beetle, (3) soil centipede, (4) larva of a ground beetle, (5) stone centipede, (6) ground beetle, (7) larva of a robber fly, (8) japygid, (9) spider, (10) predatory mite.

activities and enter a state of arrested development—a form of insect hibernation called diapause (*dia* = through; *pauein* = cessation). Many of the wasp and fly parasites have finished devouring their hosts by autumn. They pupate at the onset of winter and await the arrival of spring to transform to adults. Different adult insects such as lacewings, ladybird beetles, ground beetles, rove beetles, minute pirate bugs, and spined stink bugs are all insect predators that nestle in the leaf litter for their winter repose. While egg cases of grasshoppers and pupae of caterpillars lie beneath the frozen ground, within the egg cases of praying mantises are the embryos that represent next summer's predators of these future grasshoppers and caterpillars. Larvae of firefly beetles and pupae of soldier beetles lie dormant in the leaf litter and soil until spring, not undergoing their metamorphosis until spring or early summer. Leaving the remains of flowers and grasses in a garden provides not only enriching organic matter for the soil but also a nurturing haven in which predators, parasites, and decomposers can rest secure throughout the winter months. Among these creatures are many that feed on weeds and devour the seeds of weeds. In the spring they emerge, eager to feed and replenish nutrients lost during winter rest and to take up their new year's task of maintaining the garden's natural balance.

In the interconnected worlds of our gardens, the lives of all its plant, insect, vertebrate, and microbial residents are intertwined by sometimes obvious, sometimes subtle links. By eliminating habitat for insects and other arthropods of the leaf litter, our human obsession with tidy lawns and gardens eliminates not only a rich foraging grounds for insect-eating birds and mammals throughout the months of fall, winter, and spring but also sites for their nests and a source for their nesting materials. These untidy habitats provide nutrient-rich insect meals that help assure not only survival of migratory birds during their long journeys but also survival of nestlings during the breeding seasons of our local birds. By providing habitat for the invertebrates of the leaf litter, we simultaneously provide a welcoming environment for imperiled birds.

Compare the damage done by hungry insect herbivores to vegetables that have numerous flowers as companions to damage inflicted on vegetables in a garden far removed from flowers. Do you see some of the insects that visit the flowers for pollen and nectar also surveying the surfaces of the vegetables' leaves, flowers, and fruits for prey? A rigorously performed comparison to test this hypothesis would involve daily counts of insect herbivores, insect predators, and insect parasites. However, why not just adopt a policy of planting flowers and vegetables together? After all, each flower adds color and fragrance to the garden. Give visitors to these flowers a chance to prove their value as agents of biological control of whatever pests lurk among vegetables of the garden.

Just imagine if farmers kept patches of flowers and weeds scattered throughout their vast agricultural fields as havens for pollinators and other inhabitants. Many of the insects among the inhabitants of such weed patches are predators and parasites during their early lives as larvae. Many of the larger animals such as toads, lizards, bats, and birds would find a home where they could also contribute a large share to controlling pests of adjacent agricultural crops. Wouldn't it be far wiser and more economical to let these natural predators and parasites control insect pests of gardens and farms than to spray costly and hazardous pesticides? Such toxic chemicals indiscriminately destroy not only pests of gardens but also those benefactors of gardens—the parasites, predators, and pollinators.

Even a healthy garden is not without its share of pests; for in a healthy garden, predators and parasites are always on the lookout for a meal of pests (fig. 10.10); and pests can come in many forms: insect, microbe, weed. Predators and parasites cannot exist without pests on which they can feed. Predators and parasites descend on the garden from the air and ascend from the soil below to feed on pests. Some garden predators, such as praying mantises, ladybird beetles, and rove beetles feed on pests throughout their lives. However, others, like the velvety larvae of soldier beetles and aphid-devouring larvae of hover flies, begin life as predators but switch diets later in life.

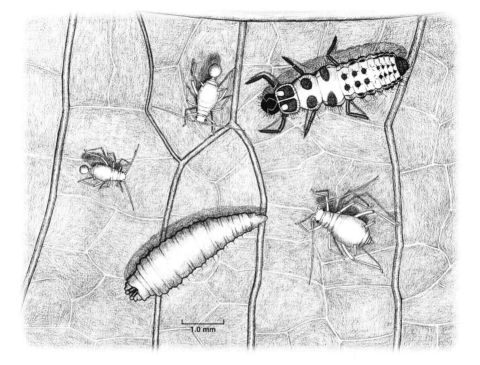

Figure 10.10 Many common insect predators aboveground provide natural biological control of aphids. The larva of a ladybird beetle approaches the aphids from the upper right, and the larva of a hover fly at lower left is about to snatch the largest of the three aphids. Both the forms and the community roles of these larval predators transform at metamorphosis. The fly and beetle parents of these larval predators are attracted to garden flowers and act as important pollinators. Being omnivores, ladybird beetle parents also diversify their diet of pollen and nectar with aphids and thrips.

As larvae, these flies and beetles are predators, and as adults they are essential pollinators. As they progress through their life cycles, these insects—like thousands of other insect species—often play not just one important role in the garden community during their brief lives, but two very different (but both important) roles. Parasitic flies and wasps feed on pests as larvae, relying on their adults to find the proper hosts for them. However, as they transform from parasitic larvae to adult flies and wasps, these insects now take on additional jobs of pollinators as they visit flowers for meals of nectar and pollen.

Multitalented fungi of the garden can also take on many jobs: as mycorrhizal partners with plant roots, assisting them in the uptake of nutrients and water; as biocontrol agents for other fungi that feed on plants; as predators that feed on nematodes and insect pests; as recyclers that are instrumental in decomposing plant matter and enriching the garden soil.

Tiny aphids, whiteflies, thrips, and flea beetles do not offer much of a meal for the birds, toads, and large praying mantises of the garden. However, if one looks closely at the leaves and flowers inhabited by these tiny insect pests, equally tiny insects that turn out to be predators and parasites of these pests are usually lurking nearby on the same leaf.

In a garden free of pesticides, all members of a balanced community are represented—herbivores, predators, parasites, decomposers. No one group becomes too abundant, and no one inflicts too much damage on the plants. As in any community, everyone has a role to play, a job to do, a task to perform.

EPILOGUE

Those who spend time in the company of plants appreciate them for their ability to sprout endless, resplendent forms from seemingly formless seeds and for the pleasures that they bring to our senses—beauty to the eye, fragrance to the nose, textures to the fingers, fine tastes to the tongue. A deeper scientific understanding of plant lives in no way diminishes the magic, mystery, and wonder we feel in the company of plants. Understanding how plants are able to accomplish these feats and impart these pleasures can actually amplify our appreciation for these fellow creatures. Indeed, the more we examine the lives of plants, the more questions arise, and certain mysteries only deepen. Science enables us to study the beauty of the plant world and to continually expand our sense of wonder.

Gardeners and farmers carry out experiments and observations each day. Whether they realize it or not, they are thinking scientifically. Close observations and constructive experiments do not always require elaborate equipment or expensive chemicals but require only that we begin to refine our perceptions of the plants that share time and space with us. In their gardens and in their laboratories, our ancestors discovered year after year new features about plant lives; and we continue to build on this body of knowledge that has been passed down to us. Curiosity drove people to ask basic questions about how plants capture the energy of light and take up nutrients from the soil or to pose more practical questions about increasing crop yields and producing sweeter carrots or redder peppers. Answers to basic questions and answers to practical questions provide not only an appreciation for what it means to be a plant but also information on helping plants make the most of their talents. Learning about the world we share with plants through observation and experiment is an adventure that amply rewards our curiosity. Thinking like a curious scientist is easy once you get the hang of it.

Many questions about the lives of plants can be asked and answered as we walk or work in the garden. We look, smell, touch, and taste as we visit the garden and listen to what our senses tell us. Observations that we gather from our senses arouse our curiosity, leading to new questioning and asking, "What if . . . ?" What if we present plants with new situations, or ask how other creatures respond to plants in such new situations? After posing a what-if hypothesis, can we predict how plants or other creatures will respond? Let your imagination be your guide.

The experiments outlined in this text are designed to test specific what-if hypotheses about plant lives. Even without attempting any of these experiments, however, we can learn from such examples about how to question plants as they carry out their everyday affairs of growing, flowering, forming fruits, and preparing for winter. With constant changes in temperature, wind, and rainfall, plants adjust to their environments aboveground and belowground. As our familiarity with the lives of plants grows, we can begin predicting how they will respond to some treatments; however, plants—like humans and other creatures—often surprise us with their responses.

When our predictions turn out to be incorrect, we are challenged anew to find an explanation, to propose a new hypothesis.

To ask what no one else has ever asked, to see what no one else has ever seen, to appreciate the singular beauty of what you see, to infer what no one else has suspected—all contribute to the delight that accompanies scientific discovery. Curiosity and imagination are crucial elements for discovery; as the father of the scientific method observed,

> They are ill discoverers that think there is no land, when they can see nothing but sea.
>
> FRANCIS BACON, 1561–1626

APPENDIX A

Important Chemicals in the Lives of Plants

Plant Hormones

Auxin

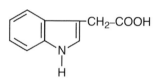

Cytokinin

Gibberellic acid

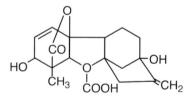

Ethylene

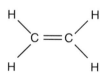

Abscisic acid

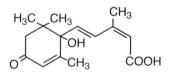

Strigolactone

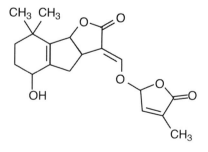

Plant Pigments

Chlorophyll

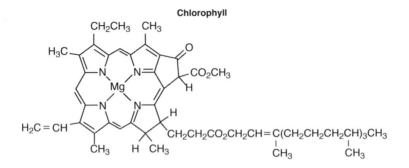

Carotenoid

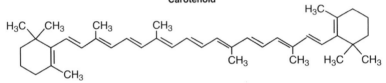

Anthocyanin

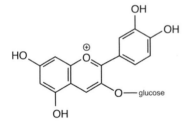

Betalain

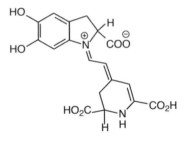

Representative Allelochemicals

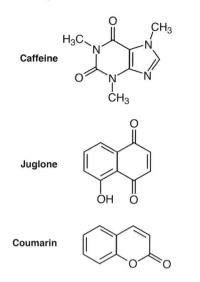

Caffeine

Juglone

Coumarin

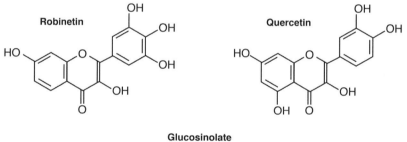

Robinetin

Quercetin

Glucosinolate

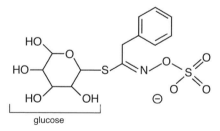

glucose

Representative Plant Defense Chemicals

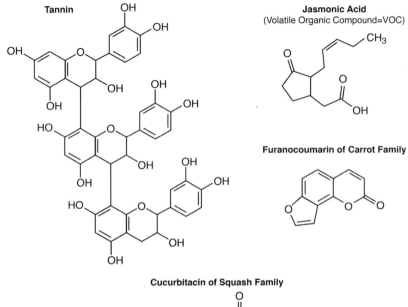

Tannin

Jasmonic Acid
(Volatile Organic Compound=VOC)

Furanocoumarin of Carrot Family

Cucurbitacin of Squash Family

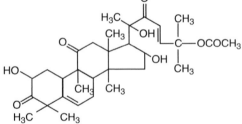

Phytoalexin of Nightshade Family

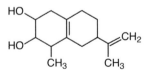

APPENDIX B

List of Plants Mentioned in the Text

Plants—vegetables, trees, fruits, garden flowers (**), weeds (*)—mentioned in this book and the families to which they belong are listed in this appendix. The plant families are listed alphabetically, and common names are listed alphabetically under each family. The genus, species, and sometimes the variety (var.) or subspecies (subsp.) of the plant follow the common names. The addition of "spp." following the common name of a plant and its genus indicates that more than one species goes by this common name.

ANGIOSPERMS, Flowering Plants

Aceraceae, maple family
  sugar maple, *Acer saccharum*
Aizoaceae, carpetweed family
  *carpetweed, *Mollugo verticillata*
Amaranthaceae, amaranth family
  *rough pigweed, *Amaranthus retroflexus*
  *smooth pigweed, *Amaranthus hybridus*
  spinach, *Spinacia oleracea*
Amaryllidaceae, amaryllis family
  **daffodil, *Narcissus* spp.

Anacardiaceae, cashew family
    *poison ivy, *Toxicodendron radicans*
    *smooth sumac, *Rhus glabra*
Apiaceae, carrot family
    carrot, *Daucus carota* var. *sativus*
    celery, *Apium graveolens*
    dill, *Anethum graveolens*
    fennel, *Foeniculum vulgare*
    parsley, *Petroselinum crispum*
    *wild carrot or Queen Anne's lace, *Daucus carota*
Araceae, arum family
    philodendron, *Philodendron* spp.
Asclepiadaceae, milkweed family
    *common milkweed, *Asclepias syriaca*
Asteraceae, daisy or aster family
    artichoke, *Cynara cardunculus* var. *scolymus*
    *aster, *Aster* spp.
    *beggar-ticks, *Bidens bipinnata*
    **black-eyed Susan, *Rudbeckia* spp.
    *burdock, *Arctium minus*
    chamomile, *Matricaria chamomilla*
    *chicory, *Cichorium intybus*
    **chrysanthemum, *Chrysanthemum* spp.
    *cocklebur, *Xanthium strumarium*
    *dandelion, *Taraxacum officinale*
    escarole and endive, *Cichorium endivia*
    *goldenrod, *Solidago canadensis*
    lettuce, *Lactuca sativa*
    marigold, *Calendula officinalis*
    **purple coneflower, *Echinacea purpurea*
    *ragweed, *Ambrosia artemisiifolia*
    sunflower, *Helianthus annuus*
    *thistle, *Cirsium vulgare*
    **zinnia, *Zinnia* spp.
Betulaceae, birch family
    birch, *Betula* spp.
Boraginaceae, borage family
    *stickseed, *Hackelia virginiana*

Brassicaceae, cabbage family

 arugula, *Eruca sativa*

 bok choy, *Brassica rapa* var. *chinensis*

 broccoli, *Brassica oleracea* var. *italic*

 Brussels sprouts, *Brassica oleracea* var. *gemmifera*

 cabbage, *Brassica oleracea* var. *capitata*

 cauliflower, *Brassica oleracea* var. *botrytis*

 Chinese cabbage, *Brassica rapa* var. *pekinensis*

 collards, *Brassica oleracea*

 kale, *Brassica oleracea*

 kohlrabi, *Brassica oleracea* var. *gongylodes*

 mustard, *Brassica juncea*

 oilseed radish, *Raphanus sativus*

 *pepperweed, *Lepidium* spp.

 radish, *Raphanus sativus*

 rutabaga, *Brassica napus*

 tatsoi, *Brassica rapa* var. *narinosa*

 turnip, *Brassica rapa*

Bromeliaceae, pineapple family

 pineapple, *Ananas comosus*

Caryophyllaceae, pink family

 *bouncing Bet, *Saponaria officinalis*

 *campion, *Lychnis* spp.

 *chickweed, *Stellaria media*

 *pink, *Silene* spp.

Chenopodiaceae, goosefoot family

 beets, *Beta vulgaris*

 *lamb's quarters, *Chenopodium album*

 Swiss chard, *Beta vulgaris*

Convolvulaceae, morning glory family

 *bindweed, *Convolvulus arvensis*

 sweet potato, *Ipomoea batatas*

Cornaceae, dogwood family

 blackgum, *Nyssa sylvatica*

 **flowering dogwood, *Cornus florida*

Cucurbitaceae, squash family

 birdhouse gourd, *Lagenaria siceraria*

 cucumber, *Cucumis sativus*

pumpkin, *Cucurbita pepo*

watermelon, *Citrullus lanatus* var. *lanatus*

zucchini, *Cucurbita pepo*

Ericaceae, heath family

blueberry, *Vaccinium corymbosum*

cranberry, *Vaccinium erythrocarpum*

Euphorbiaceae, spurge family

**poinsettia, *Euphorbia pulcherrima*

*prostrate spurge, *Euphorbia supina*

Fabaceae, pea family

alfalfa, *Medicago sativa*

beans, *Phaseolus vulgaris*

black locust, *Robinia pseudoacacia*

crimson clover, *Trifolium incarnatum*

field pea, *Pisum sativum arvense*

hairy vetch, *Vicia villosa*

pea, *Pisum sativum*

peanut, *Arachis hypogaea*

soybean, *Glycine max*

sweet clover, *Melilotus officinalis*

*tick trefoil, *Desmodium viridiflorum*

*white clover, *Trifolium repens*

Fagaceae, oak family

beech, *Fagus* spp.

red oak, *Quercus rubra*

white oak, *Quercus alba*

Geraniaceae, geranium family

*cranesbill, *Geranium carolinianum*

**cultivated geranium, *Pelargonium* spp.

**heronsbill geranium, *Erodium* spp.

*redstem storksbill or redstem filaree, *Erodium circutarium*

Hamamelidaceae, witch hazel family

sweetgum, *Liquidambar styraciflua*

witch hazel, *Hamamelis virginiana*

Hydrocharitaceae, pondweed family

waterweed, *Elodea canadensis*

Iridaceae, iris family

**crocus, *Crocus sativa*

Juglandaceae, walnut family

    black walnut, *Juglans nigra*

Lamiaceae, mint family

    basil, *Ocimum basilicum*

    catnip, *Nepeta cataria*

    **coleus, *Coleus blumei*

    oregano, *Origanum vulgare*

    peppermint, *Mentha × piperita*

    rosemary, *Rosmarinus officinalis*

    sage, *Salvia officinalis*

    summer savory, *Satureja hortensis*

    thyme, *Thymus vulgaris*

Lauraceae, laurel family

    sassafras, *Sassafras albidum*

Liliaceae, lily family

    asparagus, *Asparagus officinalis*

    chives, *Allium schoenoprasum*

    **daylily, *Hemerocallis* spp.

    garlic, *Allium sativum*

    leeks, *Allium porrum*

    onions, *Allium cepa*

Malvaceae, mallow family

    okra, *Abelmoschus esculentus*

    *spiny sida, *Sida spinosa*

    *velvetleaf, *Abutilon theophrasti*

Musaceae, banana family

    banana, *Musa* spp.

Nyctaginaceae, four o'clock family

    **four o'clock flower, *Mirabilis jalapa*

Oleaceae, olive family

    **lilac, *Syringa vulgaris*

Onagraceae, evening primrose family

    *evening primrose, *Oenothera biennis*

Orobanchaceae, broomrape family

    *witchweed, *Striga asiatica*

Oxalidaceae, wood sorrel family

    *wood sorrel, *Oxalis stricta*

Phytolaccaceae, pokeweed family

    *pokeweed, *Phytolacca americana*

Plantaginaceae, plantain family
  *broadleaf plantain, *Plantago major*
  *buckthorn plantain, *Plantago lanceolata*
Platanaceae, sycamore family
  sycamore, *Platanus* spp.
Poaceae, grass family
  barley, *Hordeum vulgare*
  big bluestem, *Andropogon gerardii*
  bluegrass, *Poa annua*
  corn, *Zea mays*
  *crabgrass, *Digitaria sanguinalis*
  *fescue, *Schedonorus phoenix*
  *foxtail, *Setaria glauca*
  oat, *Avena sativa*
  pearl millet, *Pennisetum glaucum*
  *quackgrass, *Agropyron repens*
  rye, *Secale cereale*
  sugarcane, *Saccharum officinarum*
  wheat, *Triticum aestivum*
Polygonaceae, buckwheat family
  buckwheat, *Fagopyrum esculentum*
  *curly dock, *Rumex crispus*
  *prostrate knotweed, *Polygonum aviculare*
  *sheep sorrel, *Rumex acetosella*
Portulacaceae, purslane family
  *purslane, *Portulaca oleracea*
Rosaceae, rose family
  apple, *Malus pumila*
  *avens, *Geum canadense*
  blackberry, *Rubus allegheniensis*
  pear, *Pyrus* spp.
  raspberry, *Rubus idaeus*
  **rose, *Rosa* spp.
  serviceberry, *Amelanchier* spp.
  strawberry, *Fragaria ananassa*
Rubiaceae, madder family
  *catchweed bedstraw, *Galium aparine*
  coffee, *Coffea arabica*

Salicaceae, willow family
>   cottonwood, *Populus* spp.
>   willow, *Salix* spp.

Scophulariaceae, figwort family
>   *mullein, *Verbascum thapsus*

Simaroubaceae, tree of heaven family
>   tree of heaven, *Ailanthus altissima*

Solanaceae, nightshade family
>   eggplant, *Solanum melongena*
>   pepper, *Capsicum annuum*
>   potato, *Solanum tuberosum*
>   tobacco, *Nicotiana tabacum*
>   tomato, *Solanum lycopersicum*

Violaceae, violet family
>   *common blue violet, *Viola sororia*

Vitaceae, grape family
>   table and wine grapes, *Vitis vinifera*
>   *wild grape, *Vitis vulpine*

GYMNOSPERMS, Conifers

Pinaceae, pine family
>   fir, *Abies* spp.
>   hemlock, *Tsuga* spp.
>   pine, *Pinus* spp.
>   spruce, *Picea* spp.

GLOSSARY

abscisic acid. This plant hormone influences many events in the
life of a plant, including control of germination, bud development, fruit ripening, and transpiration.

abscission (*absciss* = cut off). Separation of a leaf petiole from its
stem by disintegration of cells at the base of the petiole.

acid soil. The concentration of hydrogen ions [H^+] in acid soil is
greater than one part in ten million. In other words, its pH
($\log_{10}1/[H^+]$) is less than 7.

adventitious roots (*adventicius* = arising from outside). Rather
than sprouting from tissue derived from the hypocotyl of a
seed that is destined to become the primary root of a plant,
these roots arise from unexpected positions—such as leaves
and stems.

alkaline soil. The concentration of hydrogen ions [H^+] in alkaline
soil is less than one part in ten million. In other words, its pH
($\log_{10}1/[H^+]$) is greater than 7.

allelochemical. A naturally produced chemical that exerts an
inhibitory influence.

allelopathy (*allelo* = one another; *pathy* = harm). One plant has an
inhibitory influence on another plant.

amendments (*emendare* = improve). Certain additions to soil other than synthetic fertilizers improve its fertility and/or its structure. These amendments alter the chemical and/or physical properties of the soil.

amino acid. A building block of protein.

amyloplast (*amylo* = starch; *plast* = form). An organelle within a plant cell that stores starch and is derived from a chloroplast.

angiosperm (*angio* = enclosed; *sperm* = seed). This group of diverse vascular plants numbers around 220,000 species and produces flowers whose seeds are enclosed in fruits.

anther (*antheros* = male flower). The portion of the male (staminate) flower that holds pollen.

antheridium (*antheros* = male flower; *idium* = small). The flask-shaped structure of a fern or moss gametophyte in which sperm is formed.

anthocyanins (*anthos* = flower; *cyanos* = dark blue). A class of water-soluble red, blue, and purple pigments of plants that are localized to vacuoles of cells.

antioxidant. A chemical such as a plant pigment or vitamin that eliminates the damaging effects of other chemicals (oxidants and free radicals having unpaired electrons). The latter oxidize and chemically alter compounds in living cells through addition of oxygen and the loss of negative charge, resulting in inflammation of tissues.

apical dominance. The topmost, or apical, bud of a plant stem exerts a dominating influence on buds lower on the stem, inhibiting their growth and development.

archegonium (*archae* = primitive; *gonium* = female reproductive organ). The flask-shaped structure of fern or moss gametophyte in which eggs are formed.

arthropods (*arthro* = jointed; *poda* = leg). A large and diverse group of animals that all lack backbones and all have jointed legs. At least 1.4 million species of arthropods have been described worldwide.

autochory (*auto* = self; *chory* = dispersal). Seed dispersal carried out without assistance from animals.

auxin (*auxe* = to grow). The plant hormone that governs apical dominance of plants influences directional growth and many stages of plant development.

axil (*axilla* = armpit). The upper angle between a lateral branch, twig, or leaf petiole and the vertical axis (stem) from which it projects.

betalains. Cacti and members of the goosefoot family (beets, chard), amaranth family (spinach, pigweed), four o'clock family, and purslane family are garden plants that produce this class of yellow, orange, and red pigments. Like anthocyanins, these pigments are water soluble and localized to cell vacuoles.

bud. A structure containing stem cells located at the tip of a stem or in an axil.

bulb. Onions, leeks, and garlic are examples of vegetables whose familiar bulbs represent underground buds surrounded concentrically by scale leaves.

buzz pollination. *See* sonication pollination.

cambium (*cambium* = exchange). The meristematic layer of cells around the circumference of a stem that divides toward the outside of the stem or root (bark) to generate phloem cells and toward the interior of the stem or root to generate xylem cells.

carotenoids (*carota* = carrot). A class of plant pigments found in the membranes of chloroplasts that impart yellow and orange pigmentation.

cation. A positively charged element or nutrient.

chlorophyll (*chloro* = green; *phyll* = leaf). The green pigment of plants that captures red and blue light for photosynthesis.

chloroplast (*chloro* = green; *plast* = form). An organelle within plant cells containing chlorophyll and carotenoid pigments.

chlorosis (*chloros* = green; *osis* = diseased condition). The loss of green pigment from plants that results from a mineral deficiency.

chromoplast (*chromo* = color; *plast* = form). A chloroplast that has accumulated carotenoid pigments.

circumnutation (*circum* = around; *nuta* = nod, sway). The revolving movement of a plant part.

companion cell. The sister cell of a sieve-tube cell in phloem tissue. This cell retains all its organelles including its nucleus and supports the functioning of its sister cell that lacks a nucleus.

compost. Organic materials that have been collected and mixed outside the soil where they decompose to generate humus with minimal loss of nutrients.

compound. A chemical consisting of two or more elements combined in constant proportions.

cortex. Cells lying between (1) outermost epidermis and interior ring of endodermis in a root or (2) epidermis and vascular tissue of a stem.

cotyledon (*cotyle* = cup-shaped). This nutrient storage tissue of a seed enshrouds the plant embryo. Also known as a seed leaf.

cover crop. A crop that is planted between harvests to protect the soil from

erosion and to add mineral nutrients and organic matter to soil. Cover
crops are also known as green manure.

cucurbitacins. A class of plant defensive chemicals produced by members of
the squash family.

cytokinins. A class of plant hormones that not only promotes cell division but
also interacts with other hormones to promote growth and development
of plant tissues.

cytoskeleton. Microscopic filaments and fibers that provide internal struc-
tural support for a cell.

decomposer. An organism that obtains energy and nutrients by breaking
down the remains or waste products of other organisms.

diapause (*dia* = through; *pauein* = cessation). A period of arrested development that can occur during the life of an insect.

dicot (*di* = two; *cot* = abbreviation for cotyledon). A member of one of the
two major groups of flowering plants. A dicot seed germinates with two
cotyledons.

dinitrogenase. The enzyme of rhizobial bacteria that converts dinitrogen gas
(N_2) to ammonia (NH_3).

dormancy. A resting period associated with reduced physiological activity.

ecology (*eco* = home; *logo* = study of). The study of interactions among organisms and the interactions between organisms and their environments.

elaiosome (*elaion* = oil; *soma* = body). This nutritious protein- and oil-rich
body attached to certain seeds attracts ants that facilitate dispersal of the
seeds.

element. A chemical that cannot be broken down further to other chemicals
with different properties.

embryo. The form of a plant within a seed. Early stages of development occur
between fertilization and seed germination. The seed's embryo is often
referred to as its germ.

endodermis (*endo* = inside; *dermis* = skin). The circumferential layer of cells
in a root that selectively controls the passage of specific mineral nutrients from the soil to the central vascular system of the root.

endosperm (*endo* = within; *sperm* = seed). This nutrient storage portion of
the seed enshrouds the embryo. Endosperm originates when one of the
two sperm-cell nuclei fuses with the two nuclei (polar nuclei) of the largest of the seven cells (central cell) of the female gametophyte.

enzyme. A protein responsible for facilitating a chemical transformation.

epicotyl (*epi* = above; *cotyl* = cotyledon). The portion of the plant embryo lying above the attachment of the cotyledons and destined to form the aboveground parts of the mature plant.

epidermis (*epi* = above; *dermis* = skin). The layer of cells that covers surfaces of leaves, stems, fruits, flowers, and roots.

ethylene. This organic gas also happens to be a plant hormone that promotes leaf abscission and the ripening of fruit but that suppresses bud growth.

etiolation (*etiol* = pale). Abnormal plant growth that occurs in the absence of light and involves loss of chlorophyll, stunted leaf development, and excessive stem elongation.

fertility of soil. The ability of soil to support plants with essential nutrients for their growth.

fertilization (*fertil* = fruitful). The fusion of sperm and egg results in formation of seeds and fruit. Pollination precedes fertilization.

Fibonacci series. A series of numbers beginning with the two numbers 0 and 1, each of which is derived by adding the two previous numbers together (0, 1, 1, 2, 3, 5, 8, 13, 21 . . .). Many geometric patterns in plants, such as branching patterns and spiral arrangements of plant parts, can be described with numbers of the Fibonacci series.

fixation of carbon. The first phase of photosynthesis that combines one molecule of carbon dioxide (CO_2) with one molecule of five-carbon ribulose bisphosphate to produce two molecules of three-carbon phosphoglycerate.

fixation of nitrogen. The energy-demanding conversion of dinitrogen gas (N_2) to ammonia (NH_3).

food web. A network of interactions among organisms depicting how food energy is exchanged among plants, herbivores, predators, parasites, and decomposers.

free radical. Any negatively charged molecule with an unpaired number of electrons that can damage cells by reacting with compounds within the cells. Antioxidants donate electrons to neutralize damaging free radicals.

furanocoumarins. A class of plant defensive compounds produced by a variety of plants including members of the carrot family.

gametophyte (*gamete* = wife or husband; *phyte* = plant). The sexual phase of the plant life cycle that forms gametes (sperm + egg) and carries only half the genetic material (*n*).

germination. The sprouting of a seed or spore.

gibberellic acid. A plant hormone that promotes cell elongation, seed germination, and bud growth but inhibits leaf abscission and fruit ripening.

global warming. The warming of planet Earth attributed to an increase in atmospheric gases such as carbon dioxide, methane, and nitrous oxide that trap solar energy (heat) at Earth's surface.

glucosinolate. A simple compound that contains a glucose portion and an amino acid portion. These compounds are produced by members of the cabbage or mustard family and a few other plants. Glucosinolates have multiple functions as (1) healthy nutrients in our diet, (2) plant allelochemicals, and (3) toxic substances for most insects, but (4) enticing chemicals for a few insects.

greenhouse gas. Just as glass traps heat within a greenhouse, gases such as carbon dioxide (CO_2), methane (CH_4), and nitrous oxide (N_2O) trap heat within the atmosphere and contribute to the warming of Earth.

guttation (*gutta* = drop). The expulsion of sap droplets from special ducts at leaf tips as root pressure in xylem cells increases.

gymnosperm (*gymnos* = naked; *sperm* = seed). This group of vascular plants that numbers 720 species forms seeds that are not enclosed in fruits.

herbivore (*herbi* = plant; *vor* = eat). An organism that feeds on plants.

honeydew. Partially digested plant sap that has passed through the gut of a sap-feeding insect and that still retains some nutrients.

hormone (*hormon* = arouse). Chemical agents that regulate important events in the development of a plant.

humus (*humi* = earth). The negatively charged organic matter that remains in soil after most plant and animal matter has been decomposed.

hydathode (*hydat* = watery; *hod* = way). A duct constructed by cells at the tip of a leaf that channels water and nutrients expelled by root pressure from xylem cells.

hypocotyl (*hypo* = under; *cotyl* = cotyledon). The portion of the plant embryo lying below the attachment of the cotyledons and destined to form the underground portion of the mature plant.

hypothesis (*hypo* = beneath; *thesis* = rules). A testable explanation for a phenomenon.

jasmonic acid. A hormone that defends a plant from herbivores and influences which bacteria associate with its roots.

legume. A member of the pea family. Legumes include beans, clovers, peas, and peanuts and are known for their associations with nitrogen-fixing bacteria.

lenticel. A pore on the surface of a plant stem or potato tuber through which gases are exchanged.

manure. Animal droppings whose organic matter and inorganic nutrients enrich the soil by their addition.

mast. An exceptionally abundant harvest of fruit.

megaspore. An immature female gametophyte that divides to form the seven cells of the mature gametophyte.

meristematic (*meristos* = divisible). A region of actively dividing cells. Included among these cells are many stem cells.

metabolite (*metabol* = to change). Any chemical produced by an organism. Some of these (*primary metabolites*) are essential for the growth, development, and reproduction of the organism. Those chemicals that are important for an organism's interactions with its environment, but rarely essential for its survival, are termed *secondary metabolites*.

microbe. A creature that cannot be easily examined without a microscope. These creatures include bacteria, protozoa, fungi, and algae.

microspore. An immature pollen grain or immature male gametophyte that divides to form a mature male gametophyte or mature pollen.

mineral. An inorganic compound derived from the remains of organisms or from rocks. Some rocks such as limestone ($CaCO_3$) contain a single mineral; other rocks such as granite contain a mixture of minerals.

mitochondrion (*mitos* = thread; *chondrion* = granule). An organelle found in cells that supplies energy in the form of ATP (adenosine triphosphate).

monocot (*mono* = one; *cot* = abbreviation for cotyledon). A member of one of the two major groups of flowering plants. A monocot seed germinates with one cotyledon.

mordant (*morda* = biting). A chemical used to fix color to a fabric during the process of dyeing.

mycorrhiza (*myco* = fungus; *rhiza* = root). A mutually beneficial association of fungi with roots of plants.

myrmecochory (*myrmex* = ant; *chory* = dispersal). Seed dispersal carried out with the assistance of ants.

necrotic (*necros* = death). Localized death of tissue.

nematode (*nema* = thread; *odes* = resembling). These tiny threadworms are extremely abundant in rich soils, numbering around five million individuals per square meter. In the food web of a garden, they can feed on microbes, fungi and plant roots, and smaller nematodes. In turn, they are consumed by certain microbes, fungi, larger nematodes, and other small invertebrates.

node (*nodus* = knot). The point on a stem or rhizome from which a bud or leaf arises.

nutrient. An element or a compound that nourishes and promotes the growth of an organism.

organelle (*organ* = organ; *elle* = little). A structure within a cell with its own structure and function.

organic. A substance that is organic is derived from natural sources and always contains the elements carbon and hydrogen.

osmosis (*osmos* = push). The movement of a substance from its high concentration to its lower concentration.

ovule. The portion of the female or pistillate flower that includes the female gametophyte destined to form a plant embryo and its endosperm as well as the cells surrounding the gametophyte that are destined to form the future seed coat.

oxidation. The loss of negative charge by a compound, usually associated with the gain of an oxygen atom or the loss of a hydrogen atom.

parasite. A creature that survives at the expense of another creature known as its host. The parasite is dependent on its host and usually does not kill its host.

parasitoid. A creature that survives at the expense of its host and eventually kills its host once it matures and is no longer dependent on its host for survival.

parenchyma (*par* = beside; *enchyma* = to insert) cell. A thin-walled plant cell specialized for storage of nutrients.

parthenocarpy (*parthenos* = without fertilization, virgin; *carpy* = fruit). The development of fruit in the absence of pollination and fertilization.

pathogen. A microbe that causes disease symptoms in a plant.

petiole (*petiolus* = little stalk). The stalk that attaches a leaf blade to a stem or twig.

phloem (*phloem* = bark). Vascular tissue made up of cells that form channels for conducting the sugars produced by photosynthesis. Phloem tissue lies in a concentric ring between the outer surface of a stem, trunk, or root and the more interior vascular cells of xylem.

photoperiod. The duration and timing of light (day) and dark (night).

photorespiration (*photo* = light; *respiro* = breathe). The reversal of photosynthesis combines glucose and oxygen, converting them to carbon dioxide and water. Energy is released during photorespiration.

photosynthesis (*photo* = light; *syn* = together; *thesis* = an arranging). The process of capturing light energy with the green pigment chlorophyll and using the energy to combine carbon dioxide and water in the production of glucose and oxygen.

phytoalexin (*phyto* = plant; *alexin* = defend). A secondary metabolite induced as a defensive chemical in plant cells by an invasion of microbes.

pistil (*pistillum* = pestle). The female portion (organ) of a flower.

predator. An organism that obtains nutrients from another living organism (prey) but does not live in or on that other organism.

primary metabolite. *See* metabolite.

pollen (*pollen* = dust). A male microspore that matures and divides to form two sperm cells and one tube cell after it pollinates the pistil of a flower.

pollination. The transfer of pollen from a flower's stamen to a flower's pistil. Gymnosperms produce pollen in male cones. Pollination for gymnosperms involves transfer of pollen from male to female cones.

protozoa (*proto* = first; *zoa* = animal). These single-celled microbes include amoeboid organisms with and without shells, organisms that move by beating cilia (*cilium* = hair) that cover their surfaces, and organisms that move by waving one or more flagella (*flagellum* = whip).

reduction. The gain of negative charge by a compound, often associated with the loss of an oxygen atom or the gain of a hydrogen atom.

refraction. The bending of a light beam as it crosses the interface between one medium (usually air) and another medium (usually a liquid) is measured by an instrument called a refractometer, or Brix meter.

regenerative farming. By returning more nutrients to the soil than removed during cultivation of crops, the nutrient content and structure of soil continually improves.

rhizobium (*rhizo* = root; *bios* = life). A bacterium that lives symbiotically within the root nodules of legumes.

rhizome (*rhizo* = root). An underground stem or tuber.

root cap. A thimble-shaped layer of cells covering and protecting the root tip and its actively dividing cells. Gravity-sensing statoliths (starch grains) are found in root caps.

root hair. A projection extending into the soil from a single root epidermal cell.

salicylic acid. A hormone that induces the formation of defensive compounds in a plant and controls which microbes associate with its roots.

scale leaf. A modified leaf concentrically arranged around a bud.

sclerenchyma (*scler* = hard; *enchyma* = insert). A cell with a thick cell wall that strengthens and supports parts of plants.

secondary metabolite. *See* metabolite.

seed leaf. *See* cotyledon.

sieve-tube cell. The sister cell of a phloem companion cell. These cells are arranged in tubes. To facilitate movement of sugars and water in the sieve tubes, each cell has lost its nucleus and vacuole; all smaller organelles are restricted to the periphery of the cell. Each cell is connected to adjacent cells in its tube by perforated end walls.

sleep movement. A movement of a plant part driven by changes in turgor pressure within cells and coupled to daily light/dark cycles.

soil structure. The arrangement of soil mineral particles into naturally occurring aggregates results from the interplay of inorganic mineral particles of sand, silt, and clay with soil organic matter.

soil texture. The texture of a soil is imparted by the relative proportions of its three mineral particles of sand, silt, and clay. The three mineral particles arise from the weathering of rocks and differ in their diameters: 0.05 to 2.0 mm for sand; 0.002 to 0.05 mm for silt; < 0.002 mm for clay.

sonication pollination. The energy of sounds such as buzzing of bees can disrupt particles in or on a sample. Certain sound frequencies are required to physically dislodge pollen grains from their stamens in order to make pollen available for pollination. Also known as buzz pollination.

sporophyte (*spora* = spore; *phyte* = plant). The spore-forming phase in the life cycle of a plant that arises from the fusion of gametes (sperm + egg). Each gamete carries half the genetic material of the plant (*n*), and once the two gametes fuse during fertilization, the resulting sporophyte carries the complete genetic complement (2*n*).

stamen (*stamen* = thread). The male portion (organ) of a flower that produces pollen.

starch. Sugar (glucose) molecules joined together end to end to form a polymer of glucose that is often stored in cells during winter and converted to sugars in the spring.

statolith (*stato* = resting; *lith* = stone). A starch grain formed in a special chloroplast called an amyloplast that changes positions in response to gravity and serves as a gravity detector for plants.

stem cell. An unspecialized cell that can indefinitely divide to generate not only unspecialized, undifferentiated cells like itself but also other cells that differentiate into specialized cells.

stoma (*stoma* = mouth). A pore in the leaf surface (epidermis), each surrounded by two guard cells that expand and contract to control the size of the pore. The movement of water and gases into and out of the leaf is controlled by the opening and closing of stomata (the plural of *stoma*).

strigolactone. A hormone secreted by root cells that attracts not only beneficial symbiotic mycorrhizal fungi but also plants that are root parasites.

stylets. The mouthparts of certain insects that feed by piercing plant tissues.

subsoil. The layer of soil that lies beneath the topsoil layer and is not disrupted during cultivation.

sustainable farming. By returning to the soil the nutrients that were removed during cultivation, the nutrient content and structure of the soil is maintained and not diminished.

symbiosis (*sym* = together; *bio* = living; *sis* = the process of). An intimate, continuous, and mutually beneficial relationship between two different organisms.

systemic acquired resistance (SAR). Immunity or increased resistance to attacks from herbivores and microbes that is conferred throughout the entire plant.

tendril (*tendere* = stretch out). A modified leaf or stem that can coil around objects it contacts and provide support for the rest of the plant.

topsoil. The topmost layer of soil that is disrupted during cultivation.

totipotent cell (*toti* = all; *potent* = powerful). A cell from a given organism able to form any other cell of that organism.

tracheid (*trachea* = windpipe). A hollow, elongated, and tapering cell of xylem tissue with pitted walls. Tracheids are found in the xylem tissue of all vascular plants.

transpiration (*trans* = across; *spiro* = breathe). The emission of water vapor from pores (stomata) on leaf surfaces.

trichome (*tricho* = hair). A projection from the epidermis consisting of one or more cells. Some trichomes secrete specific substances.

tuber. A *stem tuber* is an enlarged underground stem or rhizome (e.g., potato). A *root tuber* is an enlarged storage root (e.g., sweet potato).

turgor pressure (*turgo* = swollen). The pressure exerted by water on the rigid walls of plant cells.

twiner. A plant whose apical growing tip circles around a vertical support.

vascular (*vascu* = duct). The conducting tissues of plants transport water and nutrients upward (xylem tissue) and sugars and water away from leaves (phloem tissue).

vascular plant. Any plant with conducting tissues of phloem and xylem. Such plants include all flowering plants, gymnosperms, ferns, and horsetails—but not mosses.

vessel cell. A cell of xylem tissue that is hollow, cylindrical, and open at both ends. These hollow cells line up to form long channels for conducting nutrients and water upward. Vessel cells are found only in the xylem tissue of flowering plants.

volatile organic compound (VOC). A compound expelled into the air by a
    plant after being attacked by herbivores and/or microbes.
xylem (*xylo* = wood). Vascular tissue made up of cells that conduct water and
    nutrients from the soil. Xylem cells lie between the center of the stem,
    root, or trunk and the more exterior layer of phloem vascular cells.

FURTHER READING

General

Capon, B. *Botany for Gardeners*. Portland, OR: Timber Press, 2010.

Mabey, R. *The Cabaret of Plants: Botany and the Imagination*. New York: W. W. Norton, 2016.

Martin, D. L., and K. Costello Soltys, eds. *Soil: Rodale Organic Gardening Basics*. Emmaus, PA: Rodale, 2000.

Chalker-Scott, L. *How Plants Work: The Science behind the Amazing Things Plants Do*. Portland, OR: Timber Press, 2015.

Ohlson, K. *The Soil Will Save Us*. Emmaus, PA: Rodale, 2014.

Riotte, L. *Carrots Love Tomatoes: Secrets of Companion Planting for Successful Gardening*. North Adams, MA: Storey Publishing, 1998.

Raven, P. H., R. F. Evert, and S. E. Eichhorn. *Biology of Plants*, 8th ed. W. H. Freeman, 2012.

Fellow Gardeners

Lawson, N. *The Humane Gardener: Nurturing a Backyard Habitat for Wildlife*. New York: Princeton Architectural Press, 2017.

Lowenfels, J. *Teaming with Nutrients: The Organic Gardener's Guide to Optimizing Plant Nutrition*. Portland, OR: Timber Press, 2013.

———. *Teaming with Fungi: The Organic Gardener's Guide to Mycorrhizae*. Portland, OR: Timber Press, 2017.

Lowenfels, J., and W. Lewis. *Teaming with Microbes: An Organic Gardener's Guide to the Soil Food Web*. Portland, OR: Timber Press, 2006.

Nardi, J. B. *Life in the Soil: A Guide for Naturalists and Gardeners*. Chicago: University of Chicago Press, 2007.

Pigments

Lee, D. *Nature's Palette: The Science of Plant Color*. Chicago: University of Chicago Press, 2007.

Plant Movements

Darwin, C. R. *The Movements and Habits of Climbing Plants*. London: John Murray, 1875.

———. *The Power of Movement in Plants*. With Francis Darwin. London: John Murray, 1880.

Seeds

Silvertown, J. *An Orchard Invisible: A Natural History of Seeds*. Chicago: University of Chicago Press, 2009.

Thoreau, H. D. *Faith in a Seed*. Washington, DC: Island Press, 1993.

Technical

Briggs, W. R. "How Do Sunflowers Follow the Sun—and to What End? Solar Tracking May Provide Sunflowers with an Unexpected Evolutionary Benefit." *Science* 353 (August 5, 2016): 541–42.

Cheng, F., and Z. Cheng. "Research Progress on the Use of Plant Allelopathy in Agriculture and the Physiological and Ecological Mechanisms of Allelopathy." *Frontiers in Plant Science* 6 (2015): 1020.

Conn, C. E., et al. "Convergent Evolution of Strigolactone Perception Enabled Host Detection in Parasitic Plants." *Science* 349 (July 31, 2015): 540–43.

De Vrieze, J. "The Littlest Farmhands." *Science* 349 (August 14, 2015): 680–83.

Haney, C. H., and F. M. Ausubel. "Plant Microbiome Blueprints: A Plant Defense Hormone Shapes the Root Microbiome." *Science* 349 (August 20, 2015): 788–89.

Pallardy, S. G. *Physiology of Woody Plants*. Burlington, MA: Academic Press. 2008.

Puttonen, E., C. Briese, G. Mandlburger, M. Wieser, M. Pfennigbauer, A. Zlinszky, and N. Pfeifer. "Quantification of Overnight Movement of Birch (*Betula pendula*) Branches and Foliage with Short Interval Terrestrial Laser Scanning." *Frontiers in Plant Science* 7 (February 29, 2016): 222.

Weeds

Blair, K. *The Wild Wisdom of Weeds: 13 Essential Plants for Human Survival*. White River Junction, VT: Chelsea Green, 2014.

Cocannouer, J. A. *Weeds: Guardians of the Soil*. New York: Devin-Adair, 1950.

Heiser, C. B. *Weeds in My Garden: Observations on Some Misunderstood Plants*. Portland, OR: Timber Press, 2003.

Mabey, R. *Weeds: In Defense of Nature's Most Unloved Plants*. New York: HarperCollins, 2011.

Martin, A. C. *Weeds*. New York: St. Martin's Press, 2001.

INDEX

abscisic acid: hormone interactions, 46, 104, 105f, 106; influence of, 29–30, 31f, 44; response to drought conditions, 152; structure of, 239

abscission, 30, 57, 60f

acetylsalicylic acid, 198, 199f

acidity, 161, 181, 215

adenosine triphosphate (ATP), 11, 115–17

agriculture, 6, 131–32, 231

Agropyron repens (quackgrass), 158, 159f, 248

Alaska, 111

alfalfa, 173, 174t, 246

algae, 19

alkalinity, 161, 181

allelopathy, 164–65, 173, 175, 203–4

ammonia, 181, 213, 214

amyloplasts, 12f, 24, 66–67

angiosperms, 18, 19, 20, 20f, 21, 55, 89, 97f, 243–49

anthocyanins, 11, 180, 185, 187–89, 189f, 191, 240

antioxidants, 13, 180, 187–88

aphids, 134–36, 136f, 196, 232f, 233

apical bud: on bulbs, 69–70, 71f; effect of topping, 59–61; growth-inhibiting influence of, 49–51, 51f; hypothesize, effect of a lateral bud's distance from, 50–51; observe, removal of, 48, 49f, 51f; response to light, 44–45, 45f; on tubers, 50f, 69

apple, 2f, 3f, 5f, 7f, 20f, 21, 97f, 102–6, 103f, 105f, 111, 248

arthropods, 223–24, 225f, 229f

arugula, 206f, 208f, 245

asparagus, 18, 206f, 247

aspirin, 198–200, 199f

asters, 100–102, 244

ATP (adenosine triphosphate), 11, 115–17

autochory, 166

autumn leaves: abscisic acid and, 30; abscission and, 57, 60f; colors of, 185, 187, 189; for weed control, 175, 224

auxin: chemical messaging, 45–48,
    47f, 49f; ethylene and, 57, 60f,
    105f, 106; hypothesize, effects of
    topping, 58–61; structure of, 239;
    transformation from flower to fruit
    and, 92, 94, 105f, 106
axillary buds, 48
axils, 48

Bacon, Francis, 4, 237
Balfour, Lady Eve, 129
basil, 55, 190f, 201, 206f, 247
beans, 17f, 18, 19f, 31f, 37f, 48, 49f, 113f,
    119f, 141f, 142f, 173, 174t, 206f,
    212–14, 246
beanstalks, 140–42, 141f, 142f
bedstraw (Galium), 55, 248
bee, 5f, 35f, 85f, 90, 227, 228
beeches, 111, 246
beetles, 63f, 109f, 139f, 155f, 193f, 208f,
    211f, 225f, 227, 228, 229f, 230, 232.
    See also cucumber beetles; feather-
    wing beetle; flea beetles; ground
    beetles; ladybird beetles; rove
    beetles; sap beetle
beet roots, 78f, 79
beets, 119t, 180, 181, 206f, 245
beggar-ticks (Bidens), 168, 169f, 244
Berlese funnel, 223–24, 224f
betalains, 11, 180, 185, 187–88, 191, 240
big bluestem, 116, 248
bindweed (Convolvulus), 155f, 158, 159,
    161, 245
birch trees, 147, 244
black-eyed Susan, 101–2, 244
black walnut, 204, 247
bluegrass, 65, 116, 119t, 175, 204, 248
boron, 124, 126t
Brassica oleracea, 208, 245
brassinosteroids, 46
Brix, Adolf, 137
broccoli, 119t, 188, 198, 206f, 245

Brussels sprouts, 49–50, 51f, 206f, 245
buckwheat, 174t, 248
buds: apical (see apical bud); auxin's
    chemical messaging, 45–48, 47f,
    49f; communication among cells
    and, 43–44; dividing cells and, 17;
    dominance hierarchy of, 48–50,
    49f, 50f, 51f; growth-inhibiting
    influence of apical buds, 49–51,
    51f; hormones and, 44–46; hypoth-
    esize, removal of individual buds,
    50; observe, removal of the apical
    bud, 48, 49f, 51f
bulbs, tubers, and roots: apical bud
    on bulbs, 69–70, 71f; features of
    tubers that make them stems, 69;
    functions of, 62–63; generation
    of new plants, 70–71; guttation
    process, 75–77, 76f; hypothesize,
    environmental conditions that
    promote guttation, 76–77; hypoth-
    esize, roots' ability to distinguish
    their own species, 64–65, 205;
    long-term storage of, 68; observe,
    buildup of osmotic pressure in
    roots, 75; observe, factors affect-
    ing the flow of sap, 82; observe,
    passage of nutrients and water
    from soil to plant, 77–79, 78f;
    observe, pushing of water from soil
    to leaves, 72, 74–75, 74f; observe,
    speed of root growth, 64; observe,
    staining of the starch in plant cells,
    66–67, 67f; observe, transport
    channels, 73f, 79–82, 80f; sap
    flow through xylem vessels, 73f,
    78f, 80–82, 80f, 132, 133f, 134;
    scale leaves arrangement of bulbs,
    69–70, 71f; speed of growing roots,
    64–65, 64f; transportation of
    water and nutrients, 71–72, 73f,
    77–82, 78f, 80f, 133–35; vascular

transport system, 66, 69, 70f, 73f, 77–82, 133–37

burdock (*Arctium*), 167, 169f, 244

Burroughs, John, 9

bush bean, 37f, 93f, 119t, 206f, 246

buzz pollination, 89, 90f

C_3 and C_4 plants, 115–17, 118f, 119t

cabbage, 18, 78, 111, 119t, 188, 196, 198, 203f, 206f, 207–9, 245

cabbage butterflies, 109f, 111, 196

cacti, 180

caffeine, 164–65, 204, 241

calcium, 82, 123, 126t, 161, 162

cambium ring, 37–38, 38f, 39, 40f, 133f, 135f

campions (*Lychnis*), 161, 245

carbon dioxide: observe, chemical steps in photosynthesis, 114–16; photosynthesis and, 108, 110, 111, 117, 118; stomata and, 150, 151; sustainable agriculture and, 131; virtues of weeds and, 156; in xylem vessels, 81

carbon fixation, 115

carotenoids, 10, 180, 183f, 185, 191, 240

carpetweed (*Mollugo*), 55, 56f, 243

carrots: compatibility with tomatoes, 206f, 207; fluid uptake, 40f, 78, 150; lineage, 18, 244; secondary metabolites and, 196; starch storage in, 67f

catchweed bedstraw (*Galium*), 161, 248

Cather, Willa, 2

catnip, 12f, 55, 192, 194f, 201, 202f, 247

celery, 77–79, 150, 206f, 244

chemical ecology, 163

chickweed (*Stellaria*), 161, 245

chicory, 161, 244

Chinese cabbage, 108, 109f, 245

chloride, 125

chlorophyll: colors of plants and, 178, 179, 182, 183f, 185, 187, 240; function of, 10, 12f; nutrients and, 122, 123, 124; photosynthesis and, 108, 112

chloroplasts: movement of, 181–85, 184f, 186f; observe, movement within cells, 184–85, 184f, 186f; photosynthesis and, 10

chromoplasts, 182, 183f

chrysanthemum, 102, 244

circumnutation, 152

clay, 119–21, 120f, 130f

Clements, Frederic, 161

clovers, 162, 173, 174t, 246

cobalt, 122, 125

Cocannouer, Joseph, 156

cocklebur (*Xanthium*), 119t, 167, 169f, 244

Coleus, 37f, 55, 57, 58f, 59, 183, 201, 247

colors. *See* plant colors

common knotweed, 161, 167f, 248

companion gardening, 201–3

cones of evergreen trees, 53–55, 54f

conifers, 18–20, 21

copper, 122, 124, 126t

corn: as a C_4 plant, 118f, 119t; endosperms and, 21, 22f; lineage, 18, 248; observe, leaf curling, 148, 149f; observe, sense of direction of seeds, 23–25, 24f; stomata numbers, 150

cotyledons: edible seed plants, 17f, 19–21, 20f; germination process, 17f, 18, 19f; hypothesize, effects of altering, 21–22; number of in flowering plants, 18

cover crops, 130, 173–74, 174t

crabgrass, 116, 119t, 248

cranberry, 89, 90f, 246

crane fly, 225f

cranesbills geraniums, 168, 171, 171f, 246

cucumber beetles, 139f, 196

cucurbitacins, 139f, 196, 197, 242
cytokinins, 46, 47f, 58, 60f, 61, 92, 105f, 106, 239
cytoskeleton, 11

daffodils, 99f, 243
dandelions, 17f, 63f, 116, 119t, 145, 155f, 159, 161, 166, 244
Darwin, Charles, 44, 138–40, 152
Darwin, Francis, 44, 152
decomposers, 12, 119–21, 120f, 129–31, 222–27, 225f, 233
deoxyribonucleic acid (DNA), 9
diamondback moths, 196, 207
diapause, 230
dicots, 18
dill, 196, 206f, 207, 244
dinitrogenase, 213
docks (*Rumex*), 161, 166, 248
dormancy, 27–28, 29, 30, 31f, 105f, 166, 172, 230
dyes from plants, 190–91

earthworms, 5, 63f, 211f, 222
ectomycorrhizae, 216–18, 217f, 219f
eggplant, 89, 90f, 189f, 206f, 249
elaiosomes, 166–67, 167f, 171
Elodea, 111, 184, 185, 186f, 246
Elymus repens (quackgrass), 158
Elytrigia repens (quackgrass), 158
endomycorrhizae, 218–21, 219f, 221f.
    See also vesicular-arbuscular my-
    corrhizae (VAM)
endosperm, 21, 22f, 87, 89, 218
energy and nutrients: C_3 and C_4 plants, 115–17, 118f, 119t; C_4 plants' conservation of carbon dioxide, 117; decomposers, 129–31, 222–27; distribution of nutrients up the food chain, 111; environmental impact of returning organic matter to the soil, 131–32; essential soil elements for plant growth, 122–26; extraction of matter from soil as a plant grows, 118–19; hypothesize, appeal of sugar content of plants for bugs, 136–37; hypothesize, effect of adding fertilizer, 121, 125; hypothesize, impetus for the elongation of plants, 114; hypothesize, source of a soil's fertility, 125, 127, 128f; masting behavior and, 111–12; method for measuring the sugar content of sap, 137; movement of chemical energy of sugars, 132, 133f, 134–35, 135f; nutrient deficiency symptoms, 122–25, 126t; observe, aphids ability to tap sap, 134–36, 136f; observe, C_3 and C_4 plant performance, 119t; observe, chemical steps in photosynthesis, 114–16, 116f; observe, effect of adding fertilizer, 121, 125; observe, generation of oxygen by photosynthesis, 110–11; observe, responses to exposure to light, 113–14, 113f; observe, signs of plant distress, 121; organic fertilizer, 121, 125, 130, 173–76, 222–27; oxygen production by plants, 110; photorespiration, 114; photosynthesis, 108–11, 110f, 114–16, 116f; plants' means of obtaining, 10–12, 12f; plants' sensitivity to sunlight, 112–13; results of long exposure to the sun, 111–12; sources of nutrients in soil, 119–25; synthetic fertilizer, 121, 125, 130, 214, 215; texture vs. structure of soil, 121, 130f
environmental responsibility: alternatives to synthetic additions, 161, 173–76, 204, 222–27, 231, 233; impact of returning organic matter

to the soil, 131–32, 222–27; natural
pesticides, 164, 173, 192–201; organic soil amendments and, 172–
73, 222–27; sustainable agriculture
and, 131; toll of synthetic nitrogen
fertilizer, 215
epicotyl, 18
Erodium, 168, 246
escarole, 151f, 244
essential compounds, 12–13
ethylene, 46, 57–58, 60f, 68, 104, 105f,
106, 197, 216, 239
Euphorbia (spurge), 55, 56f, 116, 117,
119t, 246

featherwing beetle, 225f
fennel, 196, 206f, 244
ferns, 19, 96f
fertilizer: hypothesize, effect of
adding, 121, 125; observe, effect of
adding, 121, 125; organic, 121, 125,
130, 173–76, 222–27; synthetic,
121, 125, 130, 214, 215
fescue, 116, 119t, 204, 248
Fibonacci series, 52–55, 54f, 55f, 56f
filaments, 11f
filarees, 168, 246
fixation solutions, 190–91
flea beetles, 111, 196, 208–9, 208f, 233
flies, 35f, 63f, 85f, 90, 139f, 193f, 211f,
225f, 227–28, 229f, 232f
flowering plants: edible seed plants,
19–21, 20f; journey to seed (*see*
journey from flower to fruit and
seed); knowing when to flower, 98–
101, 99f, 101f; life cycle of, 95, 97f,
98; movement of (*see* movement
of plants); number of cotyledons
in, 18
food web: defense mechanisms of
plants, 211–12; dual duties of garden insects, 231–33, 232f; duties

of fungi, 221, 233; habitat diversity and, 212, 230; hypothesize,
benefits of diversity in gardens,
228, 231; hypothesize, effect of
addition of organic amendments,
224, 226, 227; insect dormancy
and, 230; insects involved with,
225f, 227–28, 229f, 232f; members
of, 210; microbial partners of
plants (*see* mycorrhizae; rhizobial
bacteria); observe, recyclers, 222–
24, 225f; pesticide-free gardens,
233; recycling of nutrients, 222–
26; soil nitrogen and microbes,
226–27
four o'clocks (*Mirabilis*), 145, 180, 247
foxtail, 116, 248
free radicals, 180, 187–88. *See also*
oxidants
fruits: angiosperms and, 55, 84, 94f,
102–6; colors of, 182, 183f, 188,
189f; hypothesize, how a ripe
apple affects unripe apples, 103–4;
masting behavior of trees and, 111–
12; observe, plant pigments, 190f,
190–91; ripening of, 68, 102–6,
105f, 188–89, 189f; transition from
flower to (*see* journey from flower
to fruit and seed)
fungi: connection between trees and,
216, 217f; connection between
vegetables and, 216–18, 219f,
221f; duties of, 221, 233; hypothesize, mycorrhizal inoculants'
effects, 220; observe, relationships
between roots and soil fungi, 220,
221f; roots' control of, 218–20
furanocoumarins, 196, 244

Galium (bedstraw), 55, 248
gametophytes, 85–88, 95–98. *See also*
pollination

garlic, 69–70, 71f, 206f, 247

germination: abscisic acid and, 31f, 46, 104; cotyledons and, 18, 19f, 21–22, 22f; gibberellic acid and, 30, 31f; hormonal activity and, 30, 31, 46, 218; hypothesize, caffeine's influence on, 165; inhibition of, 164, 173, 175, 204; observing the chain of events of, 26–27, 27f; signal to germinate, 26–28, 27f, 166; temperature and, 28–29

Gibberella fujikuroi (fungus), 30

gibberellic acid: ethylene and, 57, 60f; fruit ripening and, 105f, 106; influence of, 44, 46; interactions in seeds, 30, 31f; structure of, 239; transformation from flower to fruit and, 92

glandular trichomes, 194f, 195f

global warming, 131

globular springtail, 225f

glucose, 114–15

glucosinolates, 173, 196, 197, 207, 241

golden beet, 78, 78f, 245

goldenrod, 99f, 101–2, 204, 244

gourds, 139f, 145f, 195, 195f, 245

grafting, 39–43, 41f, 42f, 101, 102

grapes, 76, 190f, 249

grapevines, 82, 249

grass clippings for weed control, 175–76

grasses, 18, 76, 116, 119t, 248

grasshoppers, 137, 211f

ground beetles, 109f, 211f, 229f, 230

growth-inhibiting factor, 29–30, 31f, 49–51, 51f

growth-promoting factor, 29–30, 31f, 45–46, 45f

guttation process, 75–77, 76f

gymnosperms, 19, 96–98, 249

herbicides, 6, 154, 156, 161, 171, 172, 204

heronsbills, 168, 246

honey bees, 5f, 35f, 85f

hormones: abscisic acid, 29–30, 31f, 44, 46, 104, 105f, 106, 152, 239; auxin, 45–48, 47f, 49f, 58–61, 60f, 92, 94, 105f, 106, 239; buds and, 44, 46, 49f; communication among cells and, 44; cytokinins, 46, 47f, 92, 239; effects on plants, 30, 31f, 45–51, 57–61, 60f, 92, 105f; ethylene, 46, 57–58, 68, 104, 105f, 106, 197, 216, 239; fruit ripening and, 104, 105f, 106; gibberellic acid, 30, 31f, 44, 46, 57, 60f, 92, 105f, 106; hypothesize, effect of hormone additives, 92, 94; interactions in seeds, 30–31, 31f; jasmonic acid, 46, 197, 216, 242; observe, how hormones affect fruit ripening, 103–6, 105f; salicylic acid, 46, 197, 199–200, 199f, 216; soil bacteria and, 216

horse manure for weed control, 162, 163, 175, 224

hover flies, 231, 232f

humus, 129, 130f

Huxley, Thomas Henry, 4

hydathodes, 76f

hypocotyl, 18, 62

insects: appeal of oils to some, 193, 195, 196, 207–9; dormancy and, 230; repellents of, 13, 164, 192–93, 195, 196, 198–99, 200–201, 202f; role in the food web (*see* food web); toxicity of glucosinolates to, 173, 196, 197

irises, 18, 246

iron, 122, 124, 126t

japygid, 229f

jasmonic acid, 46, 197, 216, 242

journey from flower to fruit and seed: contents of buds, 84, 86f; difference between seeds and spores, 94–98, 95–97f; hypothesize, flowering's reliance on exposure to light, 100–101, 101f; hypothesize, how a ripe apple affects unripe apples, 103–4; hypothesize, how hormone additives can bypass pollination and fertilization, 92, 94; hypothesize, what induces flowers to bloom, 101–2; knowing when to flower, 98–101; knowing when to ripen, 102–6, 105f; observe, how hormones affect fruit ripening, 103–6, 105f; observe, pollen's interaction with the pistil, 90–92, 93f; observe, shapes of stamens, 89–90; pollination process, 84–89, 87f, 88f, 90f

juglone, 204, 241

knotweed (*Polygonum*), 55, 139f, 248
kohlrabi, 206f, 208, 245

lacewings, 139f, 230
ladybird beetles, 230, 231, 232f
lamb's quarters (*Chenopodium*), 119t, 159, 161, 245
leaching of soil nutrients, 130, 130f, 215
leaves: detachment of aging, 57–61, 60f; differences in leaf architecture, 115, 118f; effect of excess ethylene, 57–58; effect of topping, 59–61; hypothesize, partial removal of a leaf blade's effect, 58–59; movement of (*see* movement of plants); observe, effect of topping, 61; observe, premature detachment of the petiole, 57, 58f
leeks, 69–70, 151f, 206f, 247
leghemoglobin, 213

lenticels, 69
Leopold, Aldo, viii
Lessons on Soil (Russell), 125
light: apical bud's response to, 44–45, 45f; germination and, 28; hypothesize, flowering's reliance on exposure to, 100–101, 101f; hypothesize, ultraviolet light and anthocyanin production, 188–89, 189f; observe, responses to exposure to, 113–14, 113f; plant color changes in response to, 181–82, 183f, 184f; plants' sensitivity to sunlight, 112–13; results of long exposure to the sun, 111–12; testing the response to absence or presence of, 28
lilies, 18, 204, 247
Living Soil, The (Balfour), 129
loam, 119–21
Lugol's solution, 66

Mabey, Richard, 155
macronutrients, 122–23, 126t
magnesium, 82, 123, 126t, 161, 162
malate, 115–17, 116f
manganese, 82, 122, 123, 126t
manure, 125, 162–63, 175
maple trees, 80–82, 132, 187, 189, 243
masting behavior, 111–12
McGuigan, Tony, viii, 222
melons, 18, 206f, 246
meristematic regions: buds and, 43, 50; stem cells and, 35, 36f, 37, 37f, 38f, 40f, 133f, 135f, 159
meristems, 17
methyl salicylate, 197, 198, 199–200, 199f
microbes, 120f, 129, 212–22, 223f, 226–27
micronutrients, 123–25, 126t
midge, 85f, 211f, 225f

milkweeds, 166, 244
millipede, 225f
mint, 201, 206f, 247
minute pirate bugs, 230
mitochondria, 11f, 12f
Mollugo (carpetweed), 55, 56f, 243
molybdenum, 122, 124
monocots, 18
mordant bath, 190–91
morning glories, 145, 146f, 155f, 158, 245
mosquitoes, 200–201, 202f
mosses, 19, 94–98, 95f
movement of plants: daily movements, 144–48; environmental cues, 145; hypothesize, a beanstalk's reaction to a pole, 141–42; hypothesize, accounting for leaf curling, 148, 149f; hypothesize, degree of touch needed to stimulate a response, 143–44, 144f, 145f; observe, a growing beanstalk, 140–41, 141f, 142f; observe, clasping response of a tendril, 143; observe, curling of corn leaves, 148; patterns followed by climbing plants, 138–40; response to drought conditions, 151–52; sleep movements, 146–48, 147f; tendrils and touch, 142–44, 143f, 144f; turgor pressure and, 146–48; water moving through a plant, 148–51, 151f
Movements and Habits of Climbing Plants, The (Darwin), 138
Muir, John, 92
mullein (*Verbascum*), 161, 249
mushrooms, 85f, 216, 217f
mustard oils, 196, 207, 208–9
mustard plants, 78, 127, 128f, 161, 174t, 245
mycorrhizae: connection between fungi and trees, 216, 217f; connec-

tion between fungi and vegetables, 216–18, 219f; hypothesize, mycorrhizal inoculants' effects, 220; observe, relationships between roots and soil fungi, 220, 221f; roots' control of fungi, 218–20
myrmecochory, 166

Natural History of Selborne (White), 4–5
nickel, 122, 124
nitric oxide, 215
nitrogen, 122, 125, 126t, 213–15, 226–27
nitrogen fertilizer, 162, 173, 214, 215
nitrogen fixation, 162, 173, 174t, 212–15
nitrous oxide, 215
nucleus, 10, 11f, 12f, 87f, 88f
nutrient deficiency, 121–25, 126t
nutrients. *See* energy and nutrients

oaks, 32f, 81, 111, 150, 151f, 187, 246
oats, 18, 45f, 174t, 248
odors and oils. *See* plant odors and oils
oil of wintergreen, 198
oilseed radish, 173, 174t
okra, 61, 93f, 206f, 247
On Growth and Form (Thompson), 52
onions: classification as bulbs, 69–70, 71f; companion plants for, 206f; lineage of, 18, 247; optimum long-term storage conditions, 68; pigments of, 189f, 190f
orchids, 18, 218
organelles, 10, 11f, 12f, 181, 185
organic fertilizer, 121, 125, 130, 175, 176, 224, 226, 227
oribatid mite, 120f, 225f
osmosis, 72, 73f, 74f, 137, 187
osmotic pressure, 72, 74f, 75, 82, 123, 125, 132, 187
ovule, 86, 87f, 88–89, 88f, 90, 97f, 98
Oxalis, 109f, 146, 170f, 171, 247

oxaloacetate, 115, 116f, 117
oxidants, 180, 187–88. *See also* free
     radicals
oxygen: dinitrogenase and, 213; photo-
     respiration and, 114; photosynthe-
     sis and, 108; plants' expelling of,
     110, 111; seed germination and, 28;
     stomata and, 150

palms, 18
parasites and parasitoids, 139f, 193f,
     227–33
parenchyma, 66, 67f, 79
parsley, 192, 194f, 196, 206f, 207, 244
parsnips, 68, 196
parthenocarpy, 92
pauropods, 224, 225f
peanuts, 20–21, 20f, 246
peas, 94f, 173, 174t, 212–13, 246
Pelargonium, 168, 246
peppers, 21, 22f, 93f, 180, 183f, 206f,
     221f, 249
pepperweed (*Lepidium*), 161, 167f, 245
pesticides, 6, 75, 164, 176, 231
pests. *See* food web; insects
petioles, 57, 58f
pH, 181
phloem cells: cambium ring and,
     37–38, 38f, 39, 40f; movement of
     chemical energy of sugars and, 81,
     132, 133f, 134–35, 135f; transport
     system and, 38–39, 38f, 40f, 59f,
     65–66, 73f
phosphoglycerate, 115–17
phosphorus, 121, 122, 126t, 161, 162
photorespiration, 114–17, 116f
photosynthesis: chemical steps in,
     114–16, 116f; chloroplasts and,
     10; process of, 108–11, 110f, 112,
     114–18
phytoalexins, 197, 203, 242
pickleweed, 109f, 170f, 171, 247

pigments. *See* plant colors
pigweed (*Amaranthus*), 117, 119t, 161,
     165–66, 243
pine nuts, 19
pistillate, 91
pistils, 86–92, 86f, 88f, 90f, 94f, 98. *See
     also* pollination
plantains, 161, 248
plant colors: acidity and alkalinity and,
     181; changes due to anthocyanins,
     185, 187–88; changes in response
     to light, 181–82, 184f; chlorophyll
     and, 178, 179, 182, 183f, 185, 187;
     hypothesize, chemical similarities
     and differences between plants,
     180–81; hypothesize, how chloro-
     plasts move, 185; hypothesize, the
     color a dye mixture will impart,
     191; hypothesize, ultraviolet light
     and anthocyanin production,
     188–89, 189f; observe, autumn
     colors, 185; observe, chloroplasts
     movement within cells, 184–85,
     184f, 186f; observe, dyeing fabric
     or eggs with plant pigments, 190–
     91; observe, plant colors, 180, 181f,
     182f; pigments used as dyes, 190–
     91; sources of, 178–79
plant odors and oils: ability to send
     warning odors to fellow plants,
     197–200; allelopathic properties
     of, 164, 173, 204; appeal of oils to
     some insects, 195, 195f; attractive
     and repellent properties of, 196,
     200–201; chemical responses to
     pest attacks, 196; companion gar-
     dening, 201–3; hypothesize, com-
     patibility of carrots and tomatoes,
     207; hypothesize, effect of aspirin
     on insect attacks, 198–200, 199f;
     hypothesize, flea beetles' response
     to trap crops, 207–9; hypothesize,

plant odors and oils (*continued*)
if mint is a mosquito repellent, 201, 202f; immune system-like traits of plants, 199, 203; observe, architecture of roots, 205, 207f; observe, extracting plant fragrances, 200–201; observe, influence of chemical agents on fellow plants, 198; production of, 192–93, 194f, 195f

plants: accommodating nature of, 2–3; cell structure and function, 9–10, 11f, 12f; colors of (*see* plant colors); communication with one another, 3, 197–207; community of creatures in a garden, 6 (*see also* food web); engaging in discovery about, 7–8, 13–14, 235–37; essential compounds in, 12–13; gardeners' understanding of, 4–6; means of obtaining energy and nutrients (*see* energy and nutrients); movement of (*see* movement of plants); odors and oils (*see* plant odors and oils); presentation of projects about, 8–9; roots (*see* roots); secondary metabolites function, 13; stems (*see* stems); testing hypotheses about, 4, 6; transformation from seeds (*see* seeds); unobserved lives of, 1–2

Plowman, Tim, 10
polar nuclei, 86, 88f
pole beans, 138–40, 206f, 246
pollination, 84–89, 87f, 88f, 90–92, 90f, 93f, 94f
Polygonum (knotweed), 55, 248
Portulaca (purslane), 55, 116, 117, 119t, 158–60, 160f, 166, 180, 211f, 248
potassium, 82, 121, 123, 125, 126t, 161, 162
potatoes: companion plants for, 206f; dominance hierarchy of buds,

48–50, 50f; effect of hormones on, 68, 105f, 106; flower of, 89, 90f; lineage, 41f, 249; observe, grafting results, 39–43, 41f, 42f; optimum long-term storage conditions, 68; stem classification, 69; sugar and starch storage, 65–66, 67f; vascular system, 69, 70f

Power of Movement in Plants, The (Darwin), 152
praying mantises, 85f, 230
predatory mite, 229f
Priestley, Joseph, 109–10
primrose, 166, 247
proturans, 224, 225f
pseudoscorpion, 229f
purple coneflower, 99f, 101–2, 244
purslane (*Portulaca*), 55, 116, 117, 119t, 158–60, 160f, 166, 180, 211f, 248

quackgrass (*Agropyron repens*), 116, 119t, 158, 159f, 161, 165, 166, 248

radishes, 27f, 36f, 38f, 78, 206f, 245
radish seeds, 26–27, 28, 174t
recyclers, 119, 120f, 222–24, 225f
red cabbage, 180, 181f, 182f, 245
red oak, 31–32, 32f, 246
red peppers, 180, 183f, 206f, 249
red tomatoes, 180, 183f, 206f, 249
rhizobial bacteria: diversity of bacteria in garden soil, 215–16; environmental toll of synthetic nitrogen fertilizer, 215; nitrogen-fixing bacteria, 162, 212–15, 214f; plant hormones and soil bacteria, 216; production of synthetic nitrogen fertilizer, 214
rhizobial inoculants, 212
ribulose bisphosphate, 115, 116f
Robb, Scott, 111
robber fly, 63f, 211f, 229f

roots: ability to sense up and down, 23–26, 24f, 25f; control of mycorrhizal fungi, 218–20; guttation process, 75–77, 76f; hypothesize, ability to distinguish their own species, 64–65; hypothesize, directional sense of, 23–24, 26; observe, architecture of, 205, 207f; observe, buildup of osmotic pressure in, 75; observe, passage of nutrients and water from soil to plant, 73f, 77–79, 78f; observe, pushing of water from soil to leaves, 72, 73f, 74–75, 74f; observe, relationships with soil fungi, 216–20, 217f, 219f, 221f; observe, speed of root growth, 64, 64f; observe, transport channels, 79–82, 80f, 132–35, 133f, 135f; sap flow through xylem vessels, 80–82, 80f, 132–35, 133f, 135f; speed of growing roots, 64–65, 64f; transportation of water and nutrients, 66, 69, 70f, 71–72, 73f, 77–82, 78f, 80f, 132–35, 133f, 135f

rosemary, 192, 193f, 247

roses, 76, 86f, 248

rotenone, 197

rove beetles, 63f, 229f, 230, 231

Russell, John, 125, 129

rye: as a cover crop, 127, 174t; endosperms and, 21, 22f; speed of root growth, 64–65, 64f

sage, 193f, 194f, 203f, 206f, 247

salicylic acid, 46, 197, 199–200, 199f, 216

sand, 119–21, 120f, 130f

sap: method for measuring the sugar content of sap, 137; movement through xylem vessels, 80–82; observe, aphids ability to tap sap, 134–36, 136f; observe, factors affecting the flow of sap, 81, 82

sap beetle, 225f

SAR (systemic acquired resistance), 199

scale arrangement: cones of evergreen trees, 53–55, 54f; leaves of bulbs, 69–70, 71f; observe, spiral arrangement of leaves and scales, 52–55, 53f, 54f

scientific method, 4, 237

secondary metabolites, 13, 192–93, 197, 241–42

seeds: cotyledons, 17f, 18–21, 20f, 22f; difference between spores and, 94–98, 95–97f; dormancy and, 27–32; edible seed plants, 19–21; germination and light, 28; germination and temperature, 28–29; germination signals, 26–32, 31f; growth-inhibition factor, 29–30, 31f; growth-promoting factor, 30, 31f; hormone interactions, 30–31, 31f; hypothesize, caffeine's ability to inhibit seed germination, 164–65; hypothesize, cotyledons' alterations, 21–22; hypothesize, directional sense of roots, 23–26; hypothesize, response to light, 28; journey from flower to (see journey from flower to fruit and seed); lineages of flowering plants, 18–19; nutrient storage, 21, 21f, 22f, 89, 218; observe, cotyledon structure, 18, 19f, 20f, 22f; observe, germination, 26–27, 27f; observe, sense of direction, 23–25, 24f, 25f; observe, strategies for withstanding freezing temperatures, 31–32, 32f; roots' ability to sense up and down, 23–26, 24f, 25f; spreading mechanisms of weeds, 165–68, 169f, 170f, 171f; transformation into a plant, 16–18

sheep sorrel (*Rumex*), 161, 248
sieve tubes, 133f, 134, 135f
Silene, 27, 166, 245
silt, 119–21, 120f, 130f
sleep movements, 146–48, 147f
smooth springtail, 225f
soil: content information indicated by
    weeds, 161; diversity of bacteria in
    garden soil, 215–16; environmen-
    tal impact of returning organic
    matter to the, 131–32, 224–27;
    environmental toll of synthetic
    nitrogen fertilizer, 215; essential
    elements for plant growth, 122–25,
    126t; extraction of matter from as
    a plant grows, 118–19; hypothesize,
    source of a soil's fertility, 125, 127,
    128f; nitrogen-fixing bacteria,
    162, 212–14, 214f; observe, rela-
    tionships between roots and soil
    fungi, 220, 221f; plant hormones
    and soil bacteria, 216; production
    of synthetic nitrogen fertilizer,
    214; soil bacteria and hormones,
    216; soil carbon-nitrogen ratio
    and microbes, 226–27; sources of
    nutrients in, 119–21; texture vs.
    structure of, 121, 130f
soil centipede, 63f, 229f
soldier beetles, 193f, 231
soldier fly, 225f
sonication pollination, 89, 90f
sour grass, 109f, 170f, 171, 247
spiders, 85f, 179f, 229f
spinach, 59, 59f, 93f, 119t, 180, 206f,
    243
spined stink bugs, 193f, 230
spores, 19, 94–98, 95–97f
spurge (*Euphorbia*), 55, 56f, 116, 119t,
    161, 246
squash borer, 85f, 196

squash bugs, 196
squash family, 196, 245
stamens, 86f, 89–90, 90f. *See also*
    pollination
staminate, 91
starch: conversion to sugars, 30; in
    roots, 24, 214f; sugar storage and,
    65–68, 67f, 81
statoliths, 24–26, 25f
stem cells: attributes, 34; cambium
    ring, 37–39, 38f, 39f, 40f, 133f,
    135f; hypothesize, results of graft-
    ing, 43; meristematic regions, 35,
    36f, 37, 37f; observe, grafting ex-
    periment, 39–43, 41f, 42f; observe,
    how plants grow out, 39, 40f
stems: arrangement of leaves around,
    52–55, 53f; attributes of, 34;
    features of tubers that make them
    stems, 69; hypothesize, effects of
    removing a plant branch, 55–57,
    55f, 56f; observe, spiral arrange-
    ment of leaves and scales, 52–55,
    53f, 54f
stickseed (*Hackelia*), 167, 169f, 244
stomata, 75, 81, 118f, 149–51, 151f
stone centipede, 229f
storksbills, 168, 246
strawberries, 76, 206f, 248
Striga (witchweed), 218–20, 247
strigolactones, 218–20, 239
sugar-content meter, 137
sugars: hypothesize, appeal of sugar
    content of plants for bugs, 136–37;
    method for measuring the sugar
    content of sap, 137; movement of
    chemical energy of, 132, 133f, 134–
    35, 135f; movement through xylem
    vessels, 80–82; observe, aphids
    ability to tap sap, 134–36, 136f;
    observe, factors affecting the flow

of sap, 82; plant energy and (*see* energy and nutrients); storage by root vegetables, 65–66

sulfur, 123, 126t

sunflowers, 35f, 44–45, 53f, 93f, 174t, 206f, 244

sustainable (regenerative) agriculture, 132

sweet potatoes, 65–66, 67f, 68, 69, 180, 245

Swiss chard, 179f, 180, 181f, 206f, 245

synthetic fertilizer, 121, 125, 130, 214, 215

systemic acquired resistance (SAR), 199

tannins, 59f, 197, 242

tendrils and touch, 139f, 142–44, 143f, 144f, 145f

thistles, 166, 244

Thompson, D'Arcy Wentworth, 51–52

Thoreau, Henry David, 4, 16, 160

thrips, 155, 157f, 233

thyme, 192, 193f, 201, 247

tick trefoil (*Desmodium*), 167, 246

toad, 17f, 63f, 85f, 109f, 155f, 179f, 193f, 211f, 231

tobacco seeds, 28

tomatoes: appeal of oils to some insects, 195, 198; colors of, 180, 183f; compatibility with carrots and other vegetables, 206f, 207; endosperms and, 21; flowering of, 89, 94f, 99; grafting experiment, 39–43, 41f, 42f; guttation and, 76; lineage, 18, 41f, 249; trichomes of, 195f

topping, 59–61

totipotent, 70

tracheids, 80f, 133f, 134, 135f

transpiration, 75, 149–52

transport system: movement of liquids through a plant, 71–82, 132–35, 148–51, 151f; phloem cells and, 38–39, 38f, 39f, 59f, 65–66, 73f, 132–35; roots and, 65, 66, 71–82, 73f, 78f, 80f; xylem cells and, 38, 38f, 39, 40f, 59f, 66, 71–82, 73f, 78f, 80f

trap crops, 209

trichomes, 194f, 195f

tubers. *See* bulbs, tubers, and roots

tubules, 11f, 12f

turgor pressure, 11f, 72, 146–48, 187

turnips, 67f, 78f, 164, 165, 173, 174t, 206f, 245

twiners, 139–42

ultraviolet light, 188, 189

vacuoles, 11f, 12f, 135f, 146, 185, 187, 188

van Helmont, Jan Baptista, 119

vascular transport system, 38, 38f, 40f, 66, 71–82, 73f, 80f, 132–35, 133f, 135f

vegetables: connection between fungi and, 216–18, 219f, 221f; explanations for their influence on each other, 220–22; hypothesize, how root vegetables are propagated, 68–69; sugar storage by root vegetables, 65–66

velvetleaf (*Abutilon*), 146, 247

vesicular-arbuscular mycorrhizae (VAM), 218–21, 219f, 221f. *See also* endomycorrhizae

vetches, 173, 174t, 246

vines, movement of. *See* movement of plants

violets, 17f, 155f, 159, 167f, 168, 170f, 171, 249

volatile organic compounds (VOCs), 197, 242

wasps, 35f, 109f, 139f, 227–28
waterweed (*Elodea*), 111, 184, 186f, 246
weeds: allelopathic potency of some, 164–65; control methods, 161–63, 172–76; cover crops and, 173–74, 174t; hypothesize, caffeine's ability to inhibit seed germination, 164–65; hypothesize, effectiveness of organic weed control, 172–76; hypothesize, impact of adding nutrient enrichment to soils, 162–63; nutrient value of some, 159–60; observe, dispersal strategies, 168, 169f, 170–71, 170f, 171f; opinions about, 154–55; primary and secondary metabolites and, 163–64; propagation of, 158–60, 160f; regenerative ability of, 156, 158; seed spreading mechanisms, 165–68, 167f, 169f, 170f, 171f; soil content information indicated by, 161; tilling cautions, 172; virtues of, 156, 157f, 171–72

Weeds: Guardians of the Soil (Cocannouer), 156
Weil, Simone, 13
wheat, 18, 21, 22f, 119t, 174t, 248
White, Gilbert, 4–5
whiteflies, 233
white oak, 31–32, 32f, 246
wild carrot (*Daucus*), 161, 244
willow trees, 81, 199, 249
witch hazel, 99f, 100, 246
witchweed (*Striga*), 218–20, 247
wood ashes, 162, 163
wood louse, 225f
wood sorrels (*Oxalis*), 109f, 146, 168, 170f, 171, 247

xylem cells: cambium ring and, 37–38, 38f, 39, 40f, 133f, 135f; channels, 73f, 74f, 77–82, 78f, 80f; sap flow through, 80–82, 80f; transpiration and, 149–51; transport system and, 38, 38f, 39, 39f, 40f, 59f, 66, 71–82, 73f, 133f, 135f

zinc, 122, 124, 126t
zucchini squash, 85f, 90, 91, 93f, 206f, 246